It's another great book from CGP...

GCSE Chemistry is all about **understanding how science works**.
And not only that — understanding it well enough to be able to **question**
what you hear on TV and read in the papers.

But don't panic. This book includes all the **science facts** you need to learn,
and shows you how they work in the real world. It even includes
a **free** Online Edition you can read on your computer or tablet.

How to get your free Online Edition

Just go to **cgpbooks.co.uk/extras** and enter this code...

1002 9578 1359 5647

By the way, this code only works for one person. If somebody else has used
this book before you, they might have already claimed the Online Edition.

CGP — still the best! ☺

Our sole aim here at CGP is to produce the highest
quality books — carefully written, immaculately presented
and dangerously close to being funny.

Then we work our socks off to get them
out to you — at the cheapest possible prices.

Contents

Published by CGP

From original material by Richard Parsons.

Editors:
Luke Antieul, Jane Sawers, Karen Wells.

Contributors:
Michael Aicken, Mike Bossart, Sandy Gardner, Gemma Hallam, Andy Rankin.

ISBN: 978 1 84146 640 8

With thanks to Katherine Craig for the proofreading.
With thanks to Jan Greenway, Laura Jakubowski and Laura Stoney for the copyright research.

Graph to show trend in atmospheric CO_2 concentration and global temperature on page 53 based on data by EPICA Community Members 2004 and Siegenthaler et al 2005.

Pages 78 and 87 contain public sector information published by the Health and Safety Executive and licensed under the Open Government Licence v1.0.

Groovy website: www.cgpbooks.co.uk

Printed by Elanders Ltd, Newcastle upon Tyne.
Jolly bits of clipart from CorelDRAW®

The Scientific Process

Before you get started with the really fun stuff, it's a good idea to understand exactly <u>how</u> the world of science <u>works</u>. Investigate these next few pages and you'll be laughing all day long on results day.

Scientists Come Up with <u>Hypotheses</u> — Then <u>Test</u> Them

About 100 years ago, we thought atoms looked like this.

1) Scientists try to <u>explain</u> things. Everything.

2) They start by <u>observing</u> or <u>thinking about</u> something they don't understand — it could be anything, e.g. planets in the sky, a person suffering from an illness, what matter is made of... anything.

3) Then, using what they already know (plus a bit of insight), they come up with a <u>hypothesis</u> — a possible <u>explanation</u> for what they've observed.

4) The next step is to <u>test</u> whether the hypothesis might be <u>right or not</u> — this involves <u>gathering evidence</u> (i.e. <u>data</u> from <u>investigations</u>).

5) To gather evidence the scientist uses the hypothesis to make a <u>prediction</u> — a statement based on the hypothesis that can be <u>tested</u> by carrying out <u>experiments</u>.

6) If the results from the experiments match the prediction, then the scientist can be <u>more confident</u> that the hypothesis is <u>correct</u>. This <u>doesn't</u> mean the hypothesis is <u>true</u> though — other predictions based on the hypothesis might turn out to be <u>wrong</u>.

Scientists <u>Work Together</u> to Test Hypotheses

1) Different scientists can look at the <u>same evidence</u> and interpret it in <u>different ways</u>. That's why scientists usually work in <u>teams</u> — they can share their <u>different ideas</u> on how to interpret the data they find.

2) Once a team has come up with (and tested) a hypothesis they all agree with, they'll present their work to the scientific community through <u>journals</u> and <u>scientific conferences</u> so it can be judged — this is called the <u>peer review</u> process.

Then we thought they looked like this.

3) Other scientists then <u>check</u> the team's results (by trying to <u>replicate</u> them) and carry out their own experiments to <u>collect more evidence</u>.

4) If all the experiments in the world back up the hypothesis, scientists start to have a lot of <u>confidence</u> in it. (A hypothesis that is <u>accepted</u> by pretty much every scientist is referred to as a <u>theory</u>.)

5) However, if another scientist does an experiment and the results <u>don't</u> fit with the hypothesis (and other scientists can <u>replicate</u> these results), then the hypothesis is in trouble. When this happens, scientists have to come up with a new hypothesis (maybe a <u>modification</u> of the old explanation, or maybe a completely <u>new</u> one).

Scientific Ideas <u>Change</u> as <u>New Evidence</u> is Found

And then we thought they looked like this.

1) Scientific explanations are <u>provisional</u> because they only explain the evidence that's <u>currently available</u> — new evidence may come up that can't be explained.

2) This means that scientific explanations <u>never</u> become hard and fast, totally indisputable <u>fact</u>. As <u>new evidence</u> is found (or new ways of <u>interpreting</u> existing evidence are found), hypotheses can <u>change</u> or be <u>replaced</u>.

3) Sometimes, an <u>unexpected observation</u> or <u>result</u> will suddenly throw a hypothesis into doubt and further experiments will need to be carried out. This can lead to new developments that <u>increase</u> our <u>understanding</u> of science.

<u>You expect me to believe that — then show me the evidence...</u>

If scientists think something is true, they need to produce evidence to convince others — it's all part of <u>testing a hypothesis</u>. One hypothesis might survive these tests, while others won't — it's how things progress. And along the way some hypotheses will be disproved — i.e. shown not to be true.

Quality of Data

Evidence is the key to science — but not all evidence is equally good.
The way evidence is gathered can have a big effect on how trustworthy it is...

The Bigger the Sample Size the Better

1) Data based on small samples isn't as good as data based on large samples.
A sample should be representative of the whole population (i.e. it should share as many of
the various characteristics in the population as possible) — a small sample can't do that as well.

2) The bigger the sample size the better, but scientists have to be realistic when choosing how big.
For example, if you were studying how lifestyle affects people's weight it'd be great to study everyone in
the UK (a huge sample), but it'd take ages and cost a bomb. Studying a thousand people is more realistic.

Evidence Needs to be Reliable (Repeatable and Reproducible)

Evidence is only reliable if it can be repeated (during an experiment) AND other scientists can reproduce it too
(in other experiments). If it's not reliable, you can't believe it.

RELIABLE means that the data can be repeated, and reproduced by others.

EXAMPLE: In 1989, two scientists claimed that they'd produced 'cold fusion' (the energy source of
the Sun — but without the big temperatures). It was huge news — if true, it would have meant free
energy for the world... forever. However, other scientists just couldn't reproduce the results —
so the results weren't reliable. And until they are, 'cold fusion' isn't going to be accepted as fact.

Evidence Also Needs to Be Valid

VALID means that the data is reliable AND answers the original question.

EXAMPLE: DO POWER LINES CAUSE CANCER?
Some studies have found that children who live near overhead power lines are more likely to develop
cancer. What they'd actually found was a correlation (relationship) between the variables "presence of
power lines" and "incidence of cancer" — they found that as one changed, so did the other.
But this evidence is not enough to say that the power lines cause cancer, as other explanations might
be possible. For example, power lines are often near busy roads, so the areas tested could
contain different levels of pollution from traffic. So these studies don't show a definite link
and so don't answer the original question.

Don't Always Believe What You're Being Told Straight Away

1) People who want to make a point might present data in a biased way, e.g. by overemphasising
a relationship in the data. (Sometimes without knowing they're doing it.)

2) And there are all sorts of reasons why people might want to do this — for example, companies might
want to 'big up' their products. Or make impressive safety claims.

3) If an investigation is done by a team of highly-regarded scientists it's sometimes taken more seriously
than evidence from less well known scientists.

4) But having experience, authority or a fancy qualification doesn't necessarily mean the evidence is good
— the only way to tell is to look at the evidence scientifically (e.g. is it reliable, valid, etc.).

RRRR — Remember, Reliable means Repeatable and Reproducible...

By now you should have realised how important trustworthy evidence is (even more important than a good supply
of spot cream). Without it (the evidence, not the spot cream), you just can't believe what you're being told.

Limits of Science and the Issues it Creates

Science can give us amazing things — cures for diseases, space travel, heated toilet seats...
But science has its limitations — there are questions that it just can't answer.

Some Questions Are Unanswered, Others are Unanswerable

1) Some questions are unanswered — we don't know everything and we never will. We'll find out more as new hypotheses are suggested and more experiments are done, but there'll always be stuff we don't know.

2) For example, we don't know what the exact impacts of global warming are going to be. At the moment scientists don't all agree on the answers because there isn't enough reliable and valid evidence.

3) Then there's the other type... questions that all the experiments in the world won't help us answer — the "Should we be doing this at all?" type questions. There are always two sides...

4) Think about new drugs which can be taken to boost your 'brain power'.

5) Different people have different opinions on them:

Some people think they're good... Or at least no worse than taking vitamins or eating oily fish. They could let you keep thinking for longer, or improve your memory. It's thought that new drugs could allow people to think in ways that are beyond the powers of normal brains — in effect, to become geniuses...

Other people say they're bad... taking them would give you an unfair advantage in exams, say. And perhaps people would be pressured into taking them so that they could work more effectively, and for longer hours.

6) The question of whether something is morally or ethically right or wrong can't be answered by more experiments — there is no "right" or "wrong" answer.

7) The best we can do is get a consensus from society — a judgement that most people are more or less happy to live by. Science can provide more information to help people make this judgement, and the judgement might change over time. But in the end it's up to people and their conscience.

Scientific Developments are Great, but they can Raise Issues

Scientific knowledge is increased by doing experiments. And this knowledge leads to scientific developments, e.g. new technologies or new advice. These developments can create issues though. For example:

Economic issues: Society can't always afford to do things scientists recommend (e.g. investing heavily in alternative energy sources) without cutting back elsewhere.

Social issues: Decisions based on scientific evidence affect people — e.g. should fossil fuels be taxed more highly (to invest in alternative energy)? Should alcohol be banned (to prevent health problems)? Would the effect on people's lifestyles be acceptable...

Environmental issues: Chemical fertilisers may help us produce more food — but they also cause environmental problems.

Ethical issues: There are a lot of things that scientific developments have made possible, but should we do them? E.g. clone humans, develop better nuclear weapons.

Chips or rice? — totally unanswerable by science...

Science can't tell you whether you should or shouldn't do something. That is up to you and society to decide.

Planning Investigations

The next few pages show how <u>investigations</u> should be carried out — by both <u>professional scientists</u> and <u>you</u>.

To Make an Investigation a Fair Test You Have to Control the Variables

1) In a lab experiment you usually <u>change one variable</u> and <u>measure</u> how it affects the <u>other variable</u>.

> **EXAMPLE:** you might change only the temperature of a chemical reaction and measure how this affects the rate of reaction.

2) To make it a fair test <u>everything else</u> that could affect the results should <u>stay the same</u> (otherwise you can't tell if the thing that's being changed is affecting the results or not — the data won't be reliable or valid).

> **EXAMPLE** continued: you need to keep the concentration of the reactants the same, otherwise you won't know if any change in the rate of reaction is caused by the change in temperature, or a difference in reactant concentration.

3) The variable that you <u>change</u> is called the <u>independent</u> variable.

4) The variable that's <u>measured</u> is called the <u>dependent</u> variable.

> **EXAMPLE continued:**
> Independent = temperature
> Dependent = rate of reaction
> Control = reactant concentration

5) The variables that you <u>keep the same</u> are called <u>control</u> variables.

6) Because you can't always control all the variables, you often need to use a <u>control experiment</u> — an experiment that's kept under the <u>same conditions</u> as the rest of the investigation, but doesn't have anything done to it. This is so that you can see what happens when you don't change anything at all.

The Equipment Used has to be Right for the Job

> *Accurate data is data that's close to the true value — see the next page.*

1) The measuring equipment you use has to be <u>sensitive enough</u> to accurately measure the chemicals you're using, e.g. if you need to measure out 11 ml of a liquid, you'll need to use a measuring cylinder that can measure to 1 ml, not 5 or 10 ml.

2) The <u>smallest change</u> a measuring instrument can <u>detect</u> is called its RESOLUTION. E.g. some mass balances have a resolution of 1 g and some have a resolution of 0.1 g.

3) Also, equipment needs to be <u>calibrated</u> so that your data is <u>more accurate</u>. E.g. mass balances need to be set to zero before you start weighing things.

Experiments Must be Safe

1) Part of planning an investigation is making sure that it's <u>safe</u>.

2) A <u>hazard</u> is something that can <u>potentially cause harm</u>.

3) There are lots of <u>hazards</u> you could be faced with during an investigation, e.g. <u>radiation</u>, <u>electricity</u>, <u>gas</u>, <u>chemicals</u> and <u>fire</u>.

4) You should always make sure that you <u>identify</u> all the hazards that you might encounter.

5) You should also come up with ways of <u>reducing the risks</u> from the hazards you've identified.

6) One way of doing this is to carry out a <u>risk assessment</u>:

> For an experiment involving a <u>Bunsen burner</u>, the risk assessment might be something like this:

> <u>Hazard:</u> Bunsen burner is a fire risk.
> <u>Precautions:</u>
> • Keep flammable chemicals away from the Bunsen.
> • Never leave the Bunsen unattended when lit.
> • Always turn on the yellow safety flame when not in use.

Hazard: revision boredom. Precaution: use CGP books

Labs are dangerous places — you need to know the <u>hazards</u> of what you're doing <u>before you start</u>.

Collecting Data

There are a few things that can be done to make sure that you get the <u>best results</u> you possibly can.

Data Should be as <u>Reliable, Accurate</u> and <u>Precise</u> as Possible

1) When carrying out an investigation, you can <u>improve</u> the reliability of your results (see p. 2) by <u>repeating</u> the readings and calculating the mean (average, see next page). You should repeat readings at least <u>twice</u> (so that you have at least <u>three</u> readings to calculate an average result).

2) To make sure your results are reliable you can cross check them by taking a <u>second set of readings</u> with <u>another instrument</u> (or a <u>different observer</u>).

3) Checking your results match with <u>secondary sources</u>, e.g. studies that other people have done, also increases the reliability of your data.

4) You should always make sure that your results are <u>accurate</u>. Really accurate results are those that are <u>really close</u> to the <u>true answer</u>.

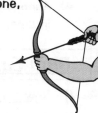

5) You can get accurate results by doing things like making sure the <u>equipment</u> you're using is <u>sensitive enough</u> (see previous page), and by recording your data to a suitable <u>level of accuracy</u>. For example, if you're taking digital readings of something, the results will be more accurate if you include at least a couple of decimal places instead of rounding to whole numbers.

6) You should also always make sure your results are <u>precise</u>. Precise results are ones where the data is <u>all really close</u> to the <u>mean</u> (i.e. not spread out).

Trial Runs <u>Help Figure out the</u> <u>Range</u> <u>and</u> <u>Interval</u> <u>of</u> <u>Variable Values</u>

1) Before you carry out an experiment, it's a good idea to do a <u>trial run</u> first — a <u>quick version</u> of your experiment.

2) Trial runs help you work out whether your plan is <u>right or not</u> — you might decide to make some <u>changes</u> after trying out your method.

3) Trial runs are used to figure out the <u>range</u> of variable values used (the upper and lower limit).

4) And they're used to figure out the <u>interval</u> (gaps) between the values too.

> Reaction example from previous page continued:
> - You might do trial runs between 10 and 60 °C. If the reaction was very slow at 10 °C and too quick to measure at 60 °C, you might narrow the range to 20-50 °C.
> - If using 10 °C intervals gives you a big change in rate of reaction you might decide to use 5 °C intervals, e.g. 20, 25, 30, 35...

5) Trial runs can also help you figure out <u>how many times</u> the experiment has to be <u>repeated</u> to get reliable results. E.g. if you repeat it two times and the <u>results</u> are all <u>similar</u>, then two repeats is enough.

<u>You Can Check For</u> <u>Mistakes Made</u> <u>When</u> <u>Collecting Data</u>

1) When you've collected all the results for an experiment, you should have a look to see if there are any results that <u>don't seem to fit</u> in with the rest.

2) Most results vary a bit, but any that are totally different are called <u>anomalous results</u>.

3) They're <u>caused</u> by <u>human errors</u>, e.g. by a whoopsie when measuring.

4) The only way to stop them happening is by taking all your measurements as <u>carefully</u> as possible.

5) If you ever get any anomalous results, you should investigate them to try to <u>work out what happened</u>. If you can work out what happened (e.g. you measured something wrong) you can <u>ignore</u> them when processing your results.

Reliable data — it won't ever forget your birthday...

All this stuff is really important — without <u>good quality</u> data an investigation will be totally <u>meaningless</u>. So give this page a read through a couple of times and your data will be the envy of the whole scientific community.

Processing, Presenting and Interpreting Data

The fun doesn't stop once the data's been collected — it then needs to be **processed** and **presented**...

Data **Needs to be** Organised

1) Data that's been collected needs to be organised so it can be processed later on.

2) Tables are dead useful for organising data.

3) When drawing tables you should always make sure that each column has a heading and that you've included the units.

4) Annoyingly, tables are about as useful as a chocolate teapot for showing patterns or relationships in data. You need to use some kind of graph or mathematical technique for that...

Test tube	Result (ml)	Repeat 1 (ml)	Repeat 2 (ml)
A	28	37	32
B	47	51	60
C	68	72	70

Data **Can be** Processed **Using a Bit of** Maths

1) Raw data generally just ain't that useful. You usually have to process it in some way.

2) A couple of the most simple calculations you can perform are the mean (average) and the range (how spread out the data is):

- To calculate the mean ADD TOGETHER all the data values and DIVIDE by the total number of values. You usually do this to get a single value from several repeats of your experiment.

- To calculate the range find the LARGEST number and SUBTRACT the SMALLEST number. You usually do this to check the accuracy and reliability of the results — the greater the spread of the data, the lower the accuracy and reliability.

Test tube	Result (ml)	Repeat 1 (ml)	Repeat 2 (ml)	Mean (ml)	Range
A	28	37	32	(28 + 37 + 32) ÷ 3 = 32.3	37 – 28 = 9
B	47	51	60	(47 + 51 + 60) ÷ 3 = 52.7	60 – 47 = 13
C	68	72	70	(68 + 72 + 70) ÷ 3 = 70.0	72 – 68 = 4

Different Types **of** Data **Should be** Presented **in** Different Ways

1) Once you've carried out an investigation, you'll need to present your data so that it's easier to see patterns and relationships in the data.

2) Different types of investigations give you different types of data, so you'll always have to choose what the best way to present your data is.

Pie charts can be used to present the same sort of data as bar charts. They're mostly used when the data is in percentages or fractions though.

Bar Charts

If the independent variable is categoric (comes in distinct categories, e.g. blood types, metals) you should use a bar chart to display the data. You also use them if the independent variable is discrete (the data can be counted in chunks, where there's no in-between value, e.g. number of people is discrete because you can't have half a person).

There are some golden rules you need to follow for drawing bar charts:

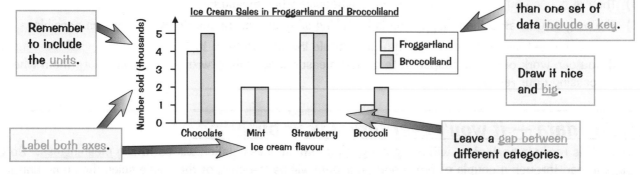

Remember to include the units.

Label both axes.

If there's more than one set of data include a key.

Draw it nice and big.

Leave a gap between different categories.

Ice Cream Sales in Froggartland and Broccoliland

Number sold (thousands)

Chocolate Mint Strawberry Broccoli
Ice cream flavour

☐ Froggartland
▨ Broccoliland

Processing, Presenting and Interpreting Data

Line Graphs

If the independent variable is <u>continuous</u> (numerical data that can have any value within a range, e.g. length, volume, temperature) you should use a <u>line graph</u> to display the data.

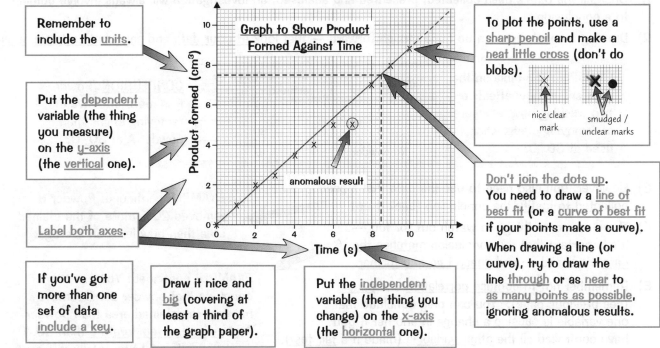

Remember to include the <u>units</u>.

Put the <u>dependent</u> variable (the thing you measure) on the <u>y-axis</u> (the <u>vertical</u> one).

Label both axes.

If you've got more than one set of data <u>include a key</u>.

Draw it nice and <u>big</u> (covering at least a third of the graph paper).

Put the <u>independent</u> variable (the thing you change) on the <u>x-axis</u> (the <u>horizontal</u> one).

Graph to Show Product Formed Against Time

anomalous result

Product formed (cm³)

Time (s)

To plot the points, use a <u>sharp pencil</u> and make a <u>neat little cross</u> (don't do blobs).

nice clear mark

smudged / unclear marks

<u>Don't join the dots up</u>. You need to draw a <u>line of best fit</u> (or a <u>curve of best fit</u> if your points make a curve).

When drawing a line (or curve), try to draw the line <u>through</u> or as <u>near</u> to <u>as many points as possible</u>, ignoring anomalous results.

Line Graphs Can Show Relationships in Data

1) Line graphs are great for showing relationships <u>between two variables</u> (just like other graphs).

2) Here are some of the different types of <u>correlation</u> (relationship) shown on line graphs:

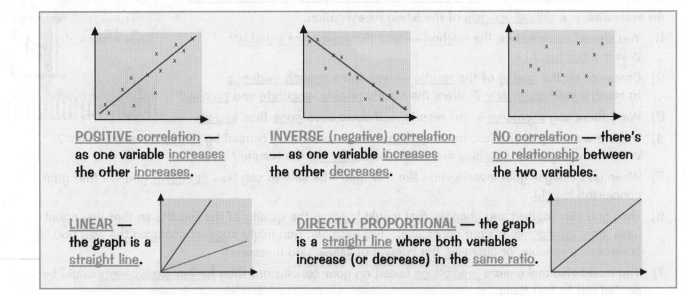

<u>POSITIVE</u> correlation — as one variable <u>increases</u> the other <u>increases</u>.

<u>INVERSE</u> (negative) correlation — as one variable <u>increases</u> the other <u>decreases</u>.

<u>NO</u> correlation — there's <u>no relationship</u> between the two variables.

<u>LINEAR</u> — the graph is a <u>straight line</u>.

<u>DIRECTLY PROPORTIONAL</u> — the graph is a <u>straight line</u> where both variables increase (or decrease) in the <u>same ratio</u>.

3) You've got to be careful not to <u>confuse correlation</u> with <u>cause</u> though. A <u>correlation</u> just means that there's a <u>relationship</u> between two variables. It <u>doesn't always mean</u> that the change in one variable is <u>causing</u> the change in the other.

4) There are <u>three possible reasons</u> for a correlation. It could be down to <u>chance</u>, it could be that there's a <u>third variable</u> linking the two things, or it might actually be that one variable is <u>causing</u> the other to change.

There's a positive correlation between age of man and length of nose hair...

<u>Process</u>, <u>present</u>, <u>interpret</u>... data's like a difficult child — it needs a lot of attention. Go on, make it happy.

Concluding and Evaluating

At the end of an investigation, the conclusion and evaluation are waiting. Don't worry, they won't bite.

A Conclusion is a Summary of What You've Learnt

1) Once all the data's been collected, presented and analysed, an investigation will always involve coming to a conclusion.

2) Drawing a conclusion can be quite straightforward — just look at your data and say what pattern you see.

EXAMPLE: The table on the right shows how effective two washing powders were at removing stains when used at 30 °C.

Washing powder	% of stained area removed on average
A	60
B	80
No powder	10

CONCLUSION: Powder B is more effective at removing stains than powder A at 30 °C.

3) However, you also need to use the data that's been collected to justify the conclusion (back it up).

EXAMPLE continued: Powder B removed 20% more of the stained area than powder A on average.

4) There are some things to watch out for too — it's important that the conclusion matches the data it's based on and doesn't go any further.

5) Remember not to confuse correlation and cause (see previous page). You can only conclude that one variable is causing a change in another if you have controlled all the other variables (made it a fair test).

EXAMPLE continued: You can't conclude that powder B will remove more of the stained area than powder A at any other temperature than 30 °C — the results might be totally different.

Evaluations — Describe How it Could be Improved

An evaluation is a critical analysis of the whole investigation.

I'd value this E somewhere in the region of 250-300k

1) You should comment on the method — was the equipment suitable? Was it a fair test?

2) Comment on the quality of the results — was there enough evidence to reach a valid conclusion? Were the results reliable, accurate and precise?

3) Were there any anomalies in the results — if there were none then say so.

4) If there were any anomalies, try to explain them — were they caused by errors in measurement? Were there any other variables that could have affected the results?

5) When you analyse your investigation like this, you'll be able to say how confident you are that your conclusion is right.

6) Then you can suggest any changes that would improve the quality of the results, so that you could have more confidence in your conclusion. For example, you might suggest changing the way you controlled a variable, or changing the interval of values you measured.

7) You could also make more predictions based on your conclusion, then further experiments could be carried out to test them.

8) When suggesting improvements to the investigation, always make sure that you say why you think this would make the results better.

Evaluation — in my next study I will make sure I don't burn the lab down...

I know it doesn't seem very nice, but writing about where you went wrong is an important skill — it shows you've got a really good understanding of what the investigation was about. It's difficult for me — I'm always right.

The History of the Atom

If you cut up a cake you end up with slices. If you keep going you're gonna have crumbs. But what happens if you keep cutting... Just how small can you go and what would the stuff you end up with look like... Scientists have been trying to work it out for years...

The Theory of Atomic Structure Has Changed Throughout History

Atoms are the tiny particles of matter (stuff that has a mass) which make up everything in the universe...

1) At the start of the 19th century John Dalton described atoms as solid spheres, and said that different spheres made up the different elements.

2) In 1897 J J Thomson concluded from his experiments that atoms weren't solid spheres. His measurements of charge and mass showed that an atom must contain even smaller, negatively charged particles — electrons. The 'solid sphere' idea of atomic structure had to be changed. The new theory was known as the 'plum pudding model'.

delicious pudding

positively charged 'pudding'

electrons

Rutherford Showed that the Plum Pudding Model Was Wrong

1) In 1909 Ernest Rutherford and his students Hans Geiger and Ernest Marsden conducted the famous gold foil experiment. They fired positively charged particles at an extremely thin sheet of gold.

2) From the plum pudding model, they were expecting most of the particles to be deflected by the positive 'pudding' that made up most of an atom. In fact, most of the particles passed straight through the gold atoms, and a very small number were deflected backwards. So the plum pudding model couldn't be right.

3) So Rutherford came up with an idea that could explain this new evidence — the theory of the nuclear atom. In this, there's a tiny, positively charged nucleus at the centre, surrounded by a 'cloud' of negative electrons — most of the atom is empty space.

A few particles are deflected backwards by the nucleus.

Most of the particles pass through empty space.

The Refined Bohr Model Explains a Lot...

1) Scientists realised that electrons in a 'cloud' around the nucleus of an atom, as Rutherford described, would be attracted to the nucleus, causing the atom to collapse. Niels Bohr proposed a new model of the atom where all the electrons were contained in shells.

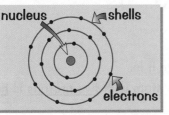

nucleus   shells

electrons

2) Bohr suggested that electrons can only exist in fixed orbits, or shells, and not anywhere in between. Each shell has a fixed energy.

3) Bohr's theory of atomic structure was supported by many experiments and it helped to explain lots of other scientists' observations at the time. It was pretty close to our currently accepted version of the atom (have a look at the next page to see what we now think atoms look like).

Scientific Theories Have to be Backed Up by Evidence

1) So, you can see that what we think the atom looks like now is completely different to what people thought in the past. These different ideas were accepted because they fitted the evidence available at the time.

2) As scientists did more experiments, new evidence was found and our theory of the structure of the atom was modified to fit it.

3) This is nearly always the way scientific knowledge develops — new evidence prompts people to come up with new, improved ideas. These ideas can be used to make predictions which if proved correct are a pretty good indication that the ideas are right.

4) Scientists also put their ideas and research up for peer review. This means everyone gets a chance to see the new ideas, check for errors and then other scientists can use it to help develop their own work.

I love a good model — Kate Moss is my favourite...

Scientists love a good theory but what they love more is trying to disprove their mate's one. That's how science works.

Atoms and Elements

Atoms are <u>amazingly tiny</u> — you can only see them with an incredibly powerful microscope.

Atoms have a Small <u>Nucleus</u> Surrounded by <u>Electrons</u>

There are quite a few different (and equally useful) modern models of the atom — but chemists tend to like this <u>model</u> best. You can use it to explain pretty much the whole of Chemistry... which is nice.

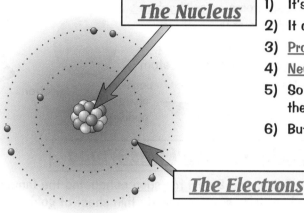

The Nucleus

1) It's in the <u>middle</u> of the atom.
2) It contains <u>protons</u> and <u>neutrons</u>.
3) <u>Protons</u> are <u>positively charged</u>.
4) <u>Neutrons</u> have <u>no charge</u> (they're neutral).
5) So the nucleus has a <u>positive charge</u> overall because of the protons.
6) But size-wise it's <u>tiny</u> compared to the rest of the atom.

The Electrons

1) Move <u>around</u> the nucleus.
2) They're <u>negatively charged</u>.
3) They're <u>tiny</u>, but they cover <u>a lot of space</u>.
4) They occupy <u>shells</u> around the nucleus.
5) These shells explain <u>the whole of Chemistry</u>.

PARTICLE	RELATIVE MASS	CHARGE
Proton	1	+1
Neutron	1	0
Electron	0.0005	-1

Number of Protons <u>Equals</u> Number of Electrons

1) Atoms have <u>no charge</u> overall. They are neutral.
2) The <u>charge</u> on the electrons is the <u>same</u> size as the charge on the <u>protons</u> — but <u>opposite</u>.
3) This means the <u>number</u> of <u>protons</u> always equals the <u>number</u> of <u>electrons</u> in an <u>atom</u>.
4) If some electrons are <u>added or removed</u>, the atom becomes <u>charged</u> and is then an <u>ion</u>.

Elements <u>Consist of One Type</u> of Atom Only

1) Atoms can have different numbers of protons, neutrons and electrons. It's the number of <u>protons</u> in the nucleus that decides what <u>type</u> of atom it is.
2) For example, an atom with <u>one proton</u> in its nucleus is <u>hydrogen</u> and an atom with <u>two protons</u> is <u>helium</u>.
3) If a substance only contains <u>one type</u> of atom it's called an <u>element</u>.
4) There are about <u>100 different elements</u> — quite a lot of everyday substances are elements:

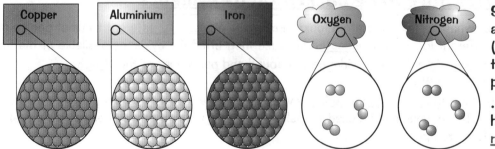

So <u>all the atoms</u> of a particular <u>element</u> (e.g. nitrogen) have the <u>same number</u> of protons...

...and <u>different elements</u> have atoms with <u>different numbers</u> of protons.

Number of protons = number of electrons...

This stuff might seem a bit useless at first, but it should be permanently engraved into your mind. <u>These basic facts</u> will give you a better chance of understanding the rest of Chemistry.

The Periodic Table

Chemistry would be <u>really messy</u> if it was all <u>big lists</u> of names and properties. So instead they've come up with a kind of <u>shorthand</u> for the names, and made a beautiful table to organise the elements — like a big <u>filing system</u>. Might not be much fun, but it makes life much, much <u>easier</u>.

Atoms Can be Represented by Symbols

Atoms of each element can be represented by a <u>one or two letter symbol</u> — it's a type of <u>shorthand</u> that saves you the bother of having to write the full name of the element.

Some make <u>perfect sense</u>, e.g.   C = carbon   O = oxygen   Mg = magnesium

Others seem to make about as much sense as an apple with a handle.

E.g.   Na = sodium   Fe = iron   Pb = lead

Most of these odd symbols actually come from the Latin names of the elements.

The Periodic Table Puts Elements with Similar Properties Together

1) The periodic table is laid out so that elements with <u>similar properties</u> form <u>columns</u>.

2) These <u>vertical columns</u> are called <u>groups</u> and Roman numerals are often used for them.

3) All of the elements in a <u>group</u> have the <u>same number</u> of <u>electrons</u> in their <u>outer shell</u>.

4) This is why <u>elements</u> in the same group have <u>similar properties</u>. So, if you know the <u>properties</u> of <u>one element</u>, you can <u>predict</u> properties of <u>other elements</u> in that group.

5) For example, the <u>Group 1</u> elements are Li, Na, K, Rb, Cs and Fr. They're all <u>metals</u> and they <u>react the same way</u>. E.g. they all react with water to form an <u>alkaline solution</u> and <u>hydrogen gas</u>, and they all react with oxygen to form an <u>oxide</u>.

6) The elements in the final column (<u>Group 0</u>) are the noble gases. They all have <u>eight electrons</u> in their <u>outer shell</u>, apart from helium (which has two). This means that they're <u>stable</u> and <u>unreactive</u>.

The top number is the <u>mass number</u>. This is the total <u>number of protons and neutrons</u>.

So, if you want to find the number of neutrons in an atom, just subtract the atomic number from the mass number.

The bottom number is the <u>atomic number</u>. This is the <u>number of protons</u>, which conveniently also tells you the <u>number of electrons</u>.

Atomic number is also known as proton number.

reactive metals | transition metals | other metals | non-metals | noble gases | separates metals from non-metals

I'm in a chemistry band — I play the symbols...

Scientists keep making <u>new elements</u> and feeling well chuffed with themselves. The trouble is, these new elements only last for <u>a fraction of a second</u> before falling apart. You can use the periodic table to estimate the properties of similar elements. If you're told, for example, that fluorine (Group 7) forms <u>two-atom molecules</u>, it's a fair guess that chlorine, bromine, iodine and astatine <u>do too</u>.

Electron Shells

The fact that electrons occupy "shells" around the nucleus is what causes the whole of chemistry. Remember that, and watch how it applies to each bit of it. It's ace.

Electron Shell Rules:

1) Electrons always occupy <u>shells</u> (sometimes called <u>energy levels</u>).

2) The <u>lowest</u> energy levels are <u>always filled first</u> — these are the ones closest to the nucleus.

3) Only <u>a certain number</u> of electrons are allowed in each shell:
   <u>1st shell:</u> 2   <u>2nd shell:</u> 8   <u>3rd shell:</u> 8

4) Atoms are much <u>happier</u> when they have <u>full electron shells</u> — like the <u>noble gases</u> in <u>Group 0</u>.

5) In most atoms the <u>outer shell</u> is <u>not full</u> and this makes the atom want to <u>react</u> to fill it.

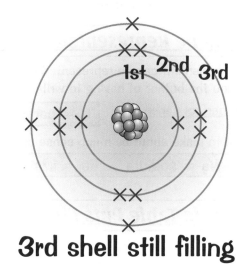

1st 2nd 3rd

3rd shell still filling

Follow the Rules to Work Out Electronic Structures

Here are the <u>electronic structures</u> for the first <u>20</u> elements (things get a bit more complicated after that). They're not too hard to work out. For a quick example, take nitrogen. <u>Follow the steps...</u>

1) The periodic table tells us nitrogen has <u>seven</u> protons... so it must have <u>seven</u> electrons.

2) Follow the 'Electron Shell Rules' above. The <u>first</u> shell can only take 2 electrons and the <u>second</u> shell can take a <u>maximum</u> of 8 electrons.

3) So the electronic structure for nitrogen <u>must</u> be <u>2, 5</u>. Easy peasy.

4) Now <u>you</u> try it for argon.

The periodic table has a big gap here where the transition metals fit in on row four.

H	Hydrogen
1	Proton no. = 1

He	Helium
2	Proton no. = 2

Li	Lithium	Be	Beryllium	B	Boron	C	Carbon	N	Nitrogen	O	Oxygen	F	Fluorine	Ne	Neon
2,1	Proton no. = 3	2,2	Proton no. = 4	2,3	Proton no. = 5	2,4	Proton no. = 6	2,5	Proton no. = 7	2,6	Proton no. = 8	2,7	Proton no. = 9	2,8	Proton no. = 10

Na	Sodium	Mg	Magnesium	Al	Aluminium	Si	Silicon	P	Phosphorus	S	Sulfur	Cl	Chlorine	Ar	Argon
2,8,1	Proton no. = 11	2,8,2	Proton no. = 12	2,8,3	Proton no. = 13	2,8,4	Proton no. = 14	2,8,5	Proton no. = 15	2,8,6	Proton no. = 16	2,8,7	Proton no. = 17	2,8,8	Proton no. = 18

K	Potassium	Ca	Calcium
2,8,8,1	Proton no. = 19	2,8,8,2	Proton no. = 20

<u>Answer...</u> To calculate the electronic structure of argon, <u>follow the rules</u>. It's got 18 protons, so it <u>must</u> have 18 electrons. The first shell must have <u>2</u> electrons, the second shell must have <u>8</u>, and so the third shell must have <u>8</u> as well. It's as easy as <u>2, 8, 8</u>.

'Electron arrangement' and 'electron configuration' are other terms for electronic structure.

One little duck and two fat ladies — 2, 8, 8...

Check you know this stuff by testing yourself until you can draw out that <u>whole diagram</u> at the bottom of the page without looking at it. Obviously, you don't have to remember each element separately, just <u>the pattern</u>. Cover the page: using a periodic table, find the atom with the electron structure 2, 8, 6.

Compounds

You can mix and match elements to make lots of compounds, which complicates things no end.

Atoms Join Together to Make Compounds

1) When different elements react, atoms form chemical bonds with other atoms to form compounds. It's usually difficult to separate the two original elements out again.

2) Making bonds involves atoms giving away, taking or sharing electrons. Only the electrons are involved — it's nothing to do with the nuclei of the atoms at all.

3) A compound which is formed from a metal and a non-metal consists of ions. The metal atoms lose electrons to form positive ions and the non-metal atoms gain electrons to form negative ions. The opposite charges (positive and negative) of the ions mean that they're strongly attracted to each other. This is called IONIC bonding.

   E.g. NaCl

   A sodium atom gives an electron to a chlorine atom.

4) A compound formed from non-metals consists of molecules. Each atom shares an electron with another atom — this is called a COVALENT bond. Each atom has to make enough covalent bonds to fill up its outer shell.

   E.g. HCl

   A hydrogen atom bonds with a chlorine atom by sharing an electron with it.

5) The properties of a compound are totally different from the properties of the original elements. For example, if iron (a lustrous magnetic metal) and sulfur (a nice yellow powder) react, the compound formed (iron sulfide) is a dull grey solid lump, and doesn't behave anything like either iron or sulfur.

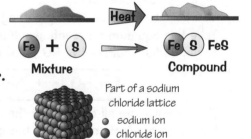

6) Compounds can be small molecules like water, or great whopping lattices like sodium chloride (when I say whopping I'm talking in atomic terms).

a water molecule

Part of a sodium chloride lattice
● sodium ion
● chloride ion

Displayed and Molecular Formulas Show the Atoms in a Substance

You can work out how many atoms of each type there are in a substance when you're given its formula. Here are some examples:

This is called a molecular formula. It shows the number and type of atoms in a molecule.

$$CH_4$$

H
H-C-H
H

Methane contains 1 carbon atom and 4 hydrogen atoms.

$$H_2O$$

H O H

Water contains 2 hydrogen atoms and 1 oxygen atom.

This is called a displayed formula. It shows the atoms and the covalent bonds in a molecule as a picture.

Don't panic if the molecular formula has brackets in it. They're easy too.

$$CH_3(CH_2)_2CH_3$$

The 2 after the bracket means that there are 2 lots of CH_2. So altogether there are 4 carbon atoms and 10 hydrogen atoms.

Drawing the displayed formula of the compound is a good way to count up the number of atoms.

Do it a bit at a time.

$$CH_3(CH_2)_2CH_3$$

H H H H
H-C-C-C-C-H
H H H H

Hopefully this page hasn't compounded your problems...

So, now you know that atoms can be very caring and sharing little things when it comes to forming compounds. Don't let yourself get confused about the difference between covalent and ionic bonding though.

Balancing Equations

Equations need a lot of practice if you're going to get them right — don't just skate over this stuff.

Atoms Aren't Lost or Made in Chemical Reactions

1) During chemical reactions, things don't appear out of nowhere and things don't just disappear.

2) You still have the same atoms at the end of a chemical reaction as you had at the start. They're just arranged in different ways.

3) Balanced symbol equations show the atoms at the start (the reactant atoms) and the atoms at the end (the product atoms) and how they're arranged. For example:

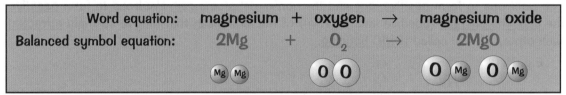

Word equation:	magnesium	+ oxygen	$\rightarrow$	magnesium oxide
Balanced symbol equation:	$2Mg$	+ O_2	$\rightarrow$	$2MgO$

4) Because atoms aren't gained or lost, the mass of the reactants equals the mass of the products. So, if you completely react 6 g of magnesium with 4 g of oxygen, you'd end up with 10 g of magnesium oxide.

Balancing the Equation — Match Them Up One by One

1) There must always be the same number of atoms of each element on both sides — they can't just disappear.

2) You balance the equation by putting numbers in front of the formulas where needed. Take this equation for reacting sulfuric acid (H_2SO_4) with sodium hydroxide (NaOH) to get sodium sulfate (Na_2SO_4) and water (H_2O):

$$H_2SO_4 + NaOH \rightarrow Na_2SO_4 + H_2O$$

The formulas are all correct but the numbers of some atoms don't match up on both sides. E.g. there are 3 Hs on the left, but only 2 on the right. You can't change formulas like H_2O to H_3O. You can only put numbers in front of them:

Method: Balance Just ONE Type of Atom at a Time

The more you practise, the quicker you get, but all you do is this:

1) Find an element that doesn't balance and pencil in a number to try and sort it out.

2) See where it gets you. It may create another imbalance — if so, just pencil in another number and see where that gets you.

3) Carry on chasing unbalanced elements and it'll sort itself out pretty quickly.

I'll show you. In the equation above you soon notice we're short of H atoms on the RHS (Right-Hand Side).

1) The only thing you can do about that is make it $2H_2O$ instead of just H_2O:
$$H_2SO_4 + NaOH \rightarrow Na_2SO_4 + 2H_2O$$

2) But that now causes too many H atoms and O atoms on the RHS, so to balance that up you could try putting 2NaOH on the LHS (Left-Hand Side):
$$H_2SO_4 + 2NaOH \rightarrow Na_2SO_4 + 2H_2O$$

3) And suddenly there it is! Everything balances. And you'll notice the Na just sorted itself out.

Balancing equations — weigh it up in your mind...

REMEMBER WHAT THOSE NUMBERS MEAN: A number in front of a formula applies to the entire formula. So, $3Na_2SO_4$ means three lots of Na_2SO_4. The little numbers in the middle or at the end of a formula only apply to the atom or brackets immediately before. So the 4 in Na_2SO_4 just means 4 Os, not 4 Ss.

Kinetic Theory and Forces Between Particles

You can explain a lot of things (including perfumes) if you get your head round this lot.

States of Matter — Depend on the Forces Between Particles

All stuff is made of particles (molecules, ions or atoms) that are constantly moving, and the forces between these particles can be weak or strong, depending on whether it's a solid, liquid or a gas.

Solids

1) There are strong forces of attraction between particles, which holds them in fixed positions in a very regular lattice arrangement.

2) The particles don't move from their positions, so all solids keep a definite shape and volume, and don't flow like liquids.

3) The particles vibrate about their positions — the hotter the solid becomes, the more they vibrate (causing solids to expand slightly when heated).

 If you heat the solid (give the particles more energy), eventually the solid will melt and become liquid.

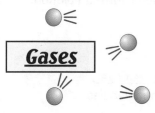

Liquids

1) There is some force of attraction between the particles. They're free to move past each other, but they do tend to stick together.

2) Liquids don't keep a definite shape and will flow to fill the bottom of a container. But they do keep the same volume.

3) The particles are constantly moving with random motion. The hotter the liquid gets, the faster they move. This causes liquids to expand slightly when heated.

 If you now heat the liquid, eventually it will boil and become gas.

Gases

1) There's next to no force of attraction between the particles — they're free to move. They travel in straight lines and only interact when they collide.

2) Gases don't keep a definite shape or volume and will always fill any container. When particles bounce off the walls of a container they exert a pressure on the walls.

3) The particles move constantly with random motion. The hotter the gas gets, the faster they move. Gases either expand when heated, or their pressure increases.

How We Smell Stuff — Volatility's the Key

1) When a liquid is heated, the heat energy goes to the particles, which makes them move faster.

2) Some particles move faster than others.

3) Fast-moving particles at the surface will overcome the forces of attraction from the other particles and escape. This is evaporation.

4) How easily a liquid evaporates is called its volatility.

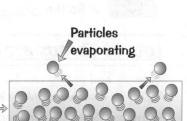

Particles evaporating

Liquid →

So... the evaporated particles are now drifting about in the air, the smell receptors in your nose pick up the chemical — and hey presto — you smell it.

Perfumes need to be quite volatile so that they can evaporate enough for you to smell them.
The particles in liquid perfumes only have a very weak attraction between them. It's easy for the particles to overcome this and escape — so you only need a very little heat energy to make the perfume evaporate.

Eau de sweaty sock — thankfully not very volatile...

Take another smelly chemical — petrol. The molecules in petrol are held together (otherwise it wouldn't be a liquid), but they must be constantly escaping (evaporating) in order for you to smell it. That's why you shouldn't have naked flames at a petrol station... the vapour from the pumps could catch fire.

Solutions

Solutions are all around you — e.g. sea water, bath salts... And inside you even — e.g. instant coffee...

A Solution is a Mixture of Solvent and Solute

When you add a solid (the solute) to a liquid (the solvent) the bonds holding the solute molecules together sometimes break and the molecules then mix with the molecules in the liquid — forming a solution. This is called dissolving. Whether the bonds break depends on how strong the attractions are between the molecules within each substance and how strong the attractions are between the two substances.

Here are some important definitions:

1) Solution – is a mixture of a solute and a solvent that does not separate out.
2) Solute – is the substance being dissolved.
3) Solvent – is the liquid it's dissolving into.
4) Soluble – means it will dissolve.
5) Insoluble – means it will NOT dissolve.
6) Solubility – a measure of how much will dissolve.

E.g. brine is a solution of salt and water — if you evaporated off the solvent (the water), you'd see the solute (the salt) again.

Water is a very common solvent.

Nail Varnish is Insoluble in Water...

Nail varnish doesn't dissolve in water. This is for two reasons:

1) The molecules of nail varnish are strongly attracted to each other. This attraction is stronger than the attraction between the nail varnish molecules and the water molecules.
2) The molecules of water are strongly attracted to each other. This attraction is stronger than the attraction between the water molecules and the nail varnish molecules.

Because the two substances are more attracted to themselves than each other, they don't form a solution.

...but Soluble in Acetone

Nail varnish dissolves in acetone — more commonly known as nail varnish remover. This is because the attraction between acetone molecules and nail varnish molecules is stronger than the attractions holding the two substances together.
So the solubility of a substance depends on the solvent used.

Acetone is also called propanone.

Lots of Things are Solvents

Alcohols and esters can be used as solvents, and so can lots of other weird and wacky organic molecules. Ability to dissolve a solute isn't the only consideration though... some solvents are horribly poisonous.

Example: Mothballs are made of a substance called naphthalene. Imagine you've trodden a mothball into your carpet. Choose one of the solvents from the table to clean it up.

Solvent	Solubility of naphthalene	Boiling point	Other properties
water	0 g/100 g	100 °C	safe
methanol	9.7 g/100 g	65 °C	flammable
ethyl acetate	18.5 g/100 g	77 °C	flammable
dichloromethane	25.0 g/100 g	40 °C	extremely toxic

Looking at the data, water wouldn't be a good choice because it doesn't dissolve the naphthalene. Dichloromethane would dissolve it easily, but it's very toxic. Of the two solvents left, ethyl acetate dissolves more naphthalene (so you won't need as much). Ethyl acetate is best (just don't set light to it).

This page has all of the solutions...

If you ever spill bright pink nail varnish (or any other colour for that matter) on your carpet, go easy with the nail varnish remover. If you use too much, the nail varnish dissolves in the remover, forming a solution which can go everywhere — and you end up with an enormous bright pink stain... aaagh.

Revision Summary for Section One

There wasn't anything too ghastly in this section, and a few bits were even quite interesting I reckon. But you've got to make sure the facts are all firmly embedded in your brain and that you really understand what's going on. These questions will let you see what you know and what you don't. If you get stuck on any, you need to look at that stuff again. Keep going till you can do them all without coming up for air.

1) Describe the famous 'gold foil experiment'. What did Rutherford conclude from it?

2) Sketch an atom. Label the nucleus and the electrons.

3) State the relative mass and charge of each particle in an atom.

4) What are the symbols for:
   a) calcium?
   b) carbon?
   c) sodium?

5)* Which element's properties are more similar to magnesium's: calcium or iron?

6) The element boron is written as $_5^{11}B$. How many neutrons does an atom of this element contain? How many electrons does a neutral boron atom have?

7) Describe how you would work out the electronic structure of an atom given its atomic number.

8) Calculate the electronic structures for each of the following elements: $_2^4He$, $_6^{12}C$, $_{15}^{31}P$, $_{19}^{39}K$.

9) Draw diagrams to show the electronic structures for the first 20 elements.

10) Describe the process of ionic bonding.

11) What is covalent bonding?

12)* A molecule has the molecular formula $CH_3(CH_2)_4CH_3$. How many H and C atoms does it contain?

13)* Write down the displayed formula for a molecule with the molecular formula C_3H_8.

14)* Balance these equations:
   a) $CaCO_3 + HCl \rightarrow CaCl_2 + H_2O + CO_2$
   b) $Ca + H_2O \rightarrow Ca(OH)_2 + H_2$

15) A substance keeps the same volume, but changes its shape according to the container it's held in. Is it a solid, a liquid or a gas? How strong are the forces of attraction between its particles?

16) What does it mean if a liquid is said to be very volatile?

17) In salt water, what is:
   a) the solute?
   b) the solution?

18) Explain why nail varnish doesn't dissolve in water.

* Answers on page 140.

Section One — Chemical Concepts

Natural and Synthetic Materials

Some materials are as natural as your birthday suit whilst others are as synthetic as a pair of PVC flares.

All Materials are Made Up of Chemicals

We use a wide range of materials, including metals, ceramics and polymers. Absolutely every material is made up of chemicals, either individual chemicals or mixtures of chemicals.

Chemicals are made up of atoms or groups of atoms bonded together called compounds (see page 13).

Some materials are mixtures of chemicals. A mixture contains different substances that are not chemically bonded together. For example, rock salt is a mixture of two compounds — salt and sand.

Some of These Materials Occur Naturally...

A lot of the materials that we use come from other living things:

MATERIALS FROM PLANTS

1) Wood and paper are both made from trees.
2) Cotton comes from the cotton plant.

MATERIALS FROM ANIMALS

1) Wool comes from sheep.
2) Silk is made by the silkworm larva.
3) Leather comes from cows.

...Others are Synthetic — Made by Humans

We also use man-made (synthetic) materials:

1) All rubber used to come from the sap of the rubber tree. We still get a lot of rubber this way (e.g. for car tyres), but you can also make rubber in a factory. The advantage of this is that you can control its properties, making it suitable for different purposes, e.g. wetsuits.

2) A lot of clothes are made of man-made fabrics like nylon or polyester. As with synthetic rubbers, the properties of synthetic fabrics can be controlled by the manufacturer — e.g. you can make fabrics that are water-proof, super-stretchy, or sparkly.

3) Most paints are mixtures of man-made chemicals. The pigment (the colouring) and the stuff that holds it all together are designed to be tough and to stop the colour fading.

The raw materials used to make synthetic materials come from the Earth's crust. E.g. aluminium and chromium are used in a lot of metal alloys.

So silk comes out of a worm's bottom then...

Okay, so now you've read this page, you should know what a material is (and that it doesn't just mean fabric). You should also have a pretty good idea of the different kinds of materials around, and where they come from. Not bad for the first page of the section. Time to cover the page and scribble down what you can remember.

Using Limestone

Limestone's often formed from sea shells, so you might not expect that it'd be useful as a building material...

Limestone *is Mainly* Calcium Carbonate

Limestone's quarried out of the ground — it's great for making into blocks for building with. Fine old buildings like cathedrals are often made purely from limestone blocks. It's pretty sturdy stuff, but don't go thinking it doesn't react with anything.

St Paul's Cathedral is made from limestone.

1) Limestone is mainly calcium carbonate — $CaCO_3$.

2) When it's heated it thermally decomposes to make calcium oxide and carbon dioxide.

> calcium carbonate → calcium oxide + carbon dioxide
> $$CaCO_{3(s)} \rightarrow CaO_{(s)} + CO_{2(g)}$$

Thermal decomposition is when one substance chemically changes into at least two new substances when it's heated.

- When magnesium, copper, zinc and sodium carbonates are heated, they decompose in the same way. E.g. magnesium carbonate → magnesium oxide + carbon dioxide (i.e. $MgCO_3 \rightarrow MgO + CO_2$)

- However, you might have difficulty doing some of these reactions in class — a Bunsen burner can't reach a high enough temperature to thermally decompose some carbonates of Group 1 metals.

3) Calcium carbonate also reacts with acid to make a calcium salt, carbon dioxide and water. E.g.:

> calcium carbonate + sulfuric acid → calcium sulfate + carbon dioxide + water
> $$CaCO_3 + H_2SO_4 \rightarrow CaSO_4 + CO_2 + H_2O$$

- The type of salt produced depends on the type of acid. For example, a reaction with hydrochloric acid would make a chloride (e.g. $CaCl_2$).

This reaction means that limestone is damaged by acid rain (see page 51).

- Other carbonates that react with acids are magnesium, copper, zinc and sodium.

Calcium Oxide **Reacts with** Water **to Produce** Calcium Hydroxide

1) When you add water to calcium oxide you get calcium hydroxide.

> calcium oxide + water ⟶ calcium hydroxide   or $CaO + H_2O \longrightarrow Ca(OH)_2$

2) Calcium hydroxide is an alkali which can be used to neutralise acidic soil in fields. Powdered limestone can be used for this too, but the advantage of calcium hydroxide is that it works much faster.

3) Calcium hydroxide can also be used in a test for carbon dioxide. If you make a solution of calcium hydroxide in water (called limewater) and bubble gas through it, the solution will turn cloudy if there's carbon dioxide in the gas. The cloudiness is caused by the formation of calcium carbonate.

> calcium hydroxide + carbon dioxide → calcium carbonate + water
> $$Ca(OH)_2 + CO_2 \rightarrow CaCO_3 + H_2O$$

Limestone is Used to Make *Other Useful Things* **Too**

1) Powdered limestone is heated in a kiln with powdered clay to make cement.

2) Cement can be mixed with sand and water to make mortar. Mortar is the stuff you stick bricks together with. You can also add calcium hydroxide to mortar.

3) Or you can mix cement with sand and aggregate (water and gravel) to make concrete.

Limestone — a sea creature's cementery...

Wow. It sounds like you can achieve pretty much anything with limestone, possibly apart from a bouncy castle. I wonder what we'd be using instead if all those sea creatures hadn't died and conveniently become rock?

Using Limestone

Using limestone ain't all hunky-dory — tearing it out of the ground and making stuff from it causes quite a few <u>problems</u>. On top of that the actual limestone products can have <u>disadvantages</u> too...

Quarrying Limestone **Makes a Right Mess of the Landscape**

Digging limestone out of the ground can cause environmental problems.

1) For a start, it makes <u>huge ugly holes</u> which permanently change the landscape.

2) <u>Quarrying</u> processes, like blasting rocks apart with explosives, make lots of <u>noise</u> and <u>dust</u> in quiet, scenic areas.

3) Quarrying <u>destroys the habitats</u> of animals and birds.

4) The limestone needs to be <u>transported away</u> from the quarry — usually in lorries. This causes more noise and pollution.

5) Waste materials produce unsightly <u>tips</u>.

Making Stuff **from Limestone Causes Pollution Too**

1) <u>Cement factories</u> make a lot of <u>dust</u>, which can cause <u>breathing problems</u> for some people.

2) <u>Energy</u> is needed to produce cement and quicklime. The energy is likely to come from burning <u>fossil fuels</u>, which causes pollution.

See pages 51-53 for more on pollution caused by burning fossil fuels.

But on the **Plus Side...**

1) Limestone provides things that people want — like <u>houses</u> and <u>roads</u>. Chemicals used in making <u>dyes</u>, <u>paints</u> and <u>medicines</u> also come from limestone.

2) Limestone products are used to <u>neutralise acidic soil</u>. Acidity in lakes and rivers caused by <u>acid rain</u> is also <u>neutralised</u> by limestone products.

3) Limestone is also used in power station chimneys to <u>neutralise sulfur dioxide</u>, which is a cause of acid rain.

4) The quarry and associated businesses provide <u>jobs</u> for people and bring more money into the <u>local economy</u>. This can lead to <u>local improvements</u> in transport, roads, recreation facilities and health.

5) Once quarrying is complete, <u>landscaping</u> and <u>restoration</u> of the area is normally required as part of the planning permission.

Limestone Products Have **Advantages and Disadvantages**

Limestone and concrete (made from cement) are used as <u>building materials</u>.
In some cases they're <u>perfect</u> for the job, but in other cases they're a bit of a compromise.

1) Limestone is <u>widely available</u> and is <u>cheaper</u> than granite or marble. It's also a fairly easy rock to <u>cut</u>.

2) Some limestone is more <u>hard-wearing</u> than marble, but it still looks <u>attractive</u>.

3) Concrete can be poured into <u>moulds</u> to make blocks or panels that can be joined together. It's a <u>very quick and cheap</u> way of constructing buildings — <u>and it shows</u>... — concrete has got to be the most <u>hideously unattractive</u> building material ever known.

4) Limestone, concrete and cement <u>don't rot</u> when they get wet like wood does. They can't be gnawed away by <u>insects</u> or <u>rodents</u> either. And to top it off, they're <u>fire-resistant</u> too.

5) Concrete <u>doesn't corrode</u> like lots of metals do. It does have a fairly <u>low tensile strength</u> though, and can crack. If it's <u>reinforced</u> with steel bars it'll be much stronger.

Tough revision here — this stuff's rock hard...

There's a <u>downside</u> to everything, including using limestone — ripping open huge quarries definitely <u>spoils the countryside</u>. But you have to find a <u>balance</u> between the environmental and ecological factors and the economic and social factors — is it worth keeping the countryside pristine if it means loads of people have nowhere to live because there's no stuff available to build houses with?

Salt

People dig great big holes to get at salt — I can live without it myself, so long as there's plenty of ketchup...

Salt *is Left Behind by* Evaporation

1) In <u>Britain</u>, salt is extracted from <u>underground deposits</u>.

2) These underground deposits were formed when <u>ancient seas</u> containing dissolved salt <u>evaporated</u>. The salt that was left behind was buried and compressed by other layers of sediment over millions of years.

3) There are massive deposits of this <u>rock salt</u> under <u>Cheshire</u> and <u>Teeside</u>.

Rock Salt *is Extracted by* Mining

<u>Rock salt</u> is a mixture of <u>salt</u> and <u>impurities</u>. It can be extracted by normal mining or by solution mining:

Normal Salt Mining Involves Physical Extraction *of Rock Salt*

1) Rock salt is drilled, blasted and dug out and brought to the surface using machinery.

2) Most rock salt obtained through this type of mining is used on <u>roads</u> to <u>stop ice forming</u>.

3) The salt can also be separated out and used to enhance the flavour in <u>food</u> or for <u>making chemicals</u>.

> 1) The salt in the mixture melts ice by lowering the freezing point of water to around –5 °C.
> 2) The sand and grit give grip on unmelted ice.

Solution Mining *is an Alternative Method of Salt Mining*

1) In solution mining <u>water</u> is <u>injected</u> into the <u>salt deposit</u> through the outer pipe (see diagram)...

2) ...which <u>dissolves</u> the salt to make a saltwater solution called <u>brine</u>.

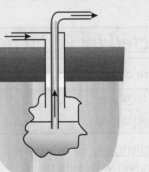

3) <u>Pressure</u> forces the brine up to the surface through the <u>inner pipe</u>.

4) The brine is then stored in <u>wells</u> above the surface and <u>pumped</u> to a <u>refining</u> plant when it's needed.

5) <u>Impurities</u> are removed from the brine in the refining plant and it's then pumped into containers. The brine is then <u>boiled</u> to make the water <u>evaporate</u>, leaving the <u>salt</u> behind.

6) Most <u>table salt</u> and a lot of the salt used for <u>chemical production</u> is produced this way.

Salt Can Also be Obtained *from the Sea*

1) In hot countries like Australia and China salt can be obtained by <u>evaporating seawater</u>.

2) Seawater flows into specially built <u>shallow pools</u> and is left to evaporate in the <u>sun</u>, leaving the salt behind.

3) This process is <u>repeated</u> several times and then the salt is <u>collected</u>.

4) This method produces the <u>purest salt</u> — it can be nearly 100% sodium chloride.

Just reading this page is making me thirsty...

Plenty of ways to get your hands on some salt then. Remember that the way salt is obtained often depends on what it's being used for — for gritting roads use rock salt. If you're gonna be eating it... sea salt is the best...

Getting Metals from Rocks

A few unreactive metals like gold are found in the Earth as the metal itself, rather than as a compound. The rest of the metals we get by extracting them from rocks — and I bet you're just itching to find out how...

Ores Contain Enough Metal to Make Extraction Worthwhile

1) A metal ore is a rock which contains enough metal to make it worthwhile extracting the metal from it.

2) In many cases the ore is an oxide of the metal. For example, the main aluminium ore is called bauxite — it's aluminium oxide (Al_2O_3).

3) Most metals need to be extracted from their ores using a chemical reaction.

4) The economics (profitability) of metal extraction can change over time. For example:

- If the market price of a metal drops a lot, it might not be worth extracting it. If the price increases a lot then it might be worth extracting more of it.
- As technology improves, it becomes possible to extract more metal from a sample of rock than was originally possible. So it might now be worth extracting metal that wasn't worth extracting in the past.

Metals Are Extracted From their Ores Chemically

1) A metal can be extracted from its ore chemically — by reduction (see below) or by electrolysis (splitting with electricity, see page 23).

2) Some ores may have to be concentrated before the metal is extracted — this just involves getting rid of the unwanted rocky material.

3) Electrolysis can also be used to purify the extracted metal (see page 23).

Occasionally, some metals are extracted from their ores using displacement reactions (see page 24).

Some Metals can be Extracted by Reduction with Carbon

1) A metal can be extracted from its ore chemically by reduction using carbon.

2) When an ore is reduced, oxygen is removed from it, e.g.

$2Fe_2O_3$	+	$3C$	$\rightarrow$	$4Fe$	+	$3CO_2$
iron(III) oxide	+	carbon	$\rightarrow$	iron	+	carbon dioxide

3) The position of the metal in the reactivity series determines whether it can be extracted by reduction with carbon.

a) Metals higher than carbon in the reactivity series have to be extracted using electrolysis, which is expensive.

b) Metals below carbon in the reactivity series can be extracted by reduction using carbon. For example, iron oxide is reduced in a blast furnace to make iron.

This is because carbon can only take the oxygen away from metals which are less reactive than carbon itself is.

Extracted using Electrolysis

Extracted by reduction using carbon

The Reactivity Series		
Potassium	K	more reactive
Sodium	Na	
Calcium	Ca	
Magnesium	Mg	
Aluminium	Al	
CARBON	C	
Zinc	Zn	
Iron	Fe	
Tin	Sn	less reactive
Copper	Cu	

Electrolysis ore reduction with carbon? — an expensive decision...

Extracting metals isn't cheap. You have to pay for special equipment, energy and labour. Then there's the cost of getting the ore to the extraction plant. If there's a choice of extraction methods, a company always picks the cheapest, unless there's a good reason not to (e.g. to increase purity). They're not extracting it for fun.

Getting Metals from Rocks

Electrolysis is an <u>expensive</u> process, but like many pricey things it's really rather good...

Some Metals *have to be* Extracted *by* Electrolysis

1) Metals that are <u>more reactive</u> than carbon (see previous page) have to be extracted using electrolysis of <u>molten compounds</u>.
2) An example of a metal that has to be extracted this way is <u>aluminium</u>.
3) However, the process is <u>much more expensive</u> than reduction with carbon (see previous page) because it <u>uses a lot of energy</u>.

> <u>FOR EXAMPLE</u>: a <u>high temperature</u> is needed to <u>melt</u> aluminium oxide so that <u>aluminium</u> can be extracted — this requires a lot of <u>energy</u>, which makes it an <u>expensive</u> process.

Copper *is Purified by* Electrolysis

1) Copper can be easily extracted by <u>reduction with carbon</u> (see previous page). The ore is <u>heated</u> in a <u>furnace</u> — this is called <u>smelting</u>.
2) However, the copper produced this way is <u>impure</u> — and impure copper <u>doesn't</u> conduct electricity very well. This <u>isn't</u> very <u>useful</u> because a lot of copper is used to make <u>electrical wiring</u>.
3) So <u>electrolysis</u> is also used to <u>purify</u> it, even though it's quite <u>expensive</u>.
4) This produces <u>very pure</u> copper, which is a <u>much better conductor</u>.

You could <u>extract</u> copper straight from its ore by electrolysis if you wanted to, but it's more expensive than using reduction with carbon.

Electrolysis Means *"Splitting Up with Electricity"*

1) <u>Electrolysis</u> is the <u>breaking down</u> of a substance using <u>electricity</u>.
2) It requires a <u>liquid</u> to <u>conduct</u> the <u>electricity</u>, called the <u>electrolyte</u>.
3) Electrolytes are often <u>metal salt solutions</u> made from the ore (e.g. copper sulfate) or <u>molten metal oxides</u>.
4) The electrolyte has <u>free ions</u> — these <u>conduct</u> the electricity and allow the whole thing to work.
5) Electrons are <u>taken away</u> by the <u>positive electrode (anode)</u> and <u>given away</u> by the <u>negative electrode (cathode)</u>. As ions gain or lose electrons they become atoms or molecules and are released.

> Here's how electrolysis is used to get <u>copper</u>:
> 1) <u>Electrons</u> are <u>pulled off</u> copper atoms at the <u>anode</u>, causing them to go into solution as Cu^{2+} ions.
> 2) Cu^{2+} ions near the <u>cathode</u> gain electrons and turn back into <u>copper atoms</u>.
> 3) The <u>impurities</u> are dropped at the <u>anode</u> as a <u>sludge</u>, whilst <u>pure copper atoms</u> bond to the <u>cathode</u>.

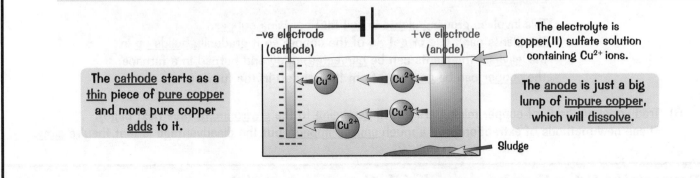

Someone robbed your metal? — call a copper...

The skin of the <u>Statue of Liberty</u> is made of copper — about 80 tonnes of it in fact. Its surface reacts with gases in the air to form <u>copper carbonate</u> — which is why it's that pretty shade of <u>green</u>.

Getting Metals from Rocks

You could use plain old iron to get hold of copper — or you could opt for newfangled fancy phytomining...

You Can Extract Copper From a Solution Using a Displacement Reaction

1) More reactive metals react more vigorously than less reactive metals.

2) If you put a reactive metal into a solution of a dissolved metal compound, the reactive metal will replace the less reactive metal in the compound.

3) This is because the more reactive metal bonds more strongly to the non-metal bit of the compound and pushes out the less reactive metal.

4) For example, scrap iron can be used to displace copper from solution — this is really useful because iron is cheap but copper is expensive. If some iron is put in a solution of copper sulfate, the more reactive iron will "kick out" the less reactive copper from the solution. You end up with iron sulfate solution and copper metal.

copper sulfate + iron → iron sulfate + copper

5) If a piece of silver metal is put into a solution of copper sulfate, nothing happens. The more reactive metal (copper) is already in the solution.

Copper-rich Ores are in Short Supply

1) The supply of copper-rich ores is limited, so it's important to recycle as much copper as possible.

2) The demand for copper is growing and this may lead to shortages in the future.

3) Scientists are looking into new ways of extracting copper from low-grade ores (ores that only contain small amounts of copper) or from the waste that is currently produced when copper is extracted.

4) Examples of new methods to extract copper are bioleaching and phytomining:

Bioleaching

This uses bacteria to separate copper from copper sulfide. The bacteria get energy from the bond between copper and sulfur, separating out the copper from the ore in the process. The leachate (the solution produced by the process) contains copper, which can be extracted, e.g. by filtering.

Phytomining

This involves growing plants in soil that contains copper.
The plants can't use or get rid of the copper so it gradually builds up in the leaves. The plants can be harvested, dried and burned in a furnace.
The copper can be collected from the ash left in the furnace.

5) Traditional methods of copper mining are pretty damaging to the environment (see next page). These new methods of extraction have a much smaller impact, but the disadvantage is that they're slow.

Personally, I'd rather be pound rich than copper rich...

Pure copper is expensive but exceptionally useful stuff. Just think where we'd be without good quality copper wire to conduct electricity (hmmm... how would I live without my electric pineapple corer). The fact that copper-rich ore supplies are dwindling means that scientists have to come up with ever-more-cunning methods to extract it. It so happens that these methods are also better for our lovely planet. Peace.

Impacts of Extracting Metals

Once metals are finished with, it's better to recycle them than to dig up more ore and extract fresh metal.

Mining can be Good or Bad

1) Ores are finite resources. This means that there's a <u>limited amount</u> of them — eventually, they'll run out.

2) People have to balance the <u>social</u>, <u>economic</u> and <u>environmental</u> effects of mining the ores.

3) So, mining metal ores is <u>good</u> because <u>useful products</u> can be made. It also provides local people with <u>jobs</u> and brings <u>money</u> into the area. This means services such as <u>transport</u> and <u>health</u> can be improved.

4) But mining ores is <u>bad for the environment</u> as it uses loads of energy, scars the landscape and destroys habitats. Also, noise, dust and pollution are caused by an increase in traffic.

5) Deep mine shafts can also be <u>dangerous</u> for a long time after the mine has been abandoned.

6) Land above disused mines can <u>collapse into the holes</u> — this is called <u>subsidence</u>. Subsidence can affect buildings near the mines, including people's homes.

7) The risk of subsidence is <u>reduced</u> by leaving <u>well-supported</u> caverns in mines (e.g. supported by pillars of rock). Also, caverns can be spaced well apart and <u>filled in</u> when no longer used.

Recycling Metals is Important

1) Mining and extracting metals takes lots of <u>energy</u>, most of which comes from burning <u>fossil fuels</u>.

2) Fossil fuels are <u>running out</u> so it's important to <u>conserve</u> them. Not only this, but burning them contributes to <u>acid rain</u>, <u>global dimming</u> and <u>climate change</u>.

3) Recycling metals only uses a <u>small fraction</u> of the energy needed to mine and extract new metal. E.g. recycling copper only takes 15% of the energy that's needed to mine and extract new copper.

4) Energy doesn't come cheap, so recycling <u>saves money</u> too.

5) As there's a <u>finite amount</u> of each <u>metal</u> in the Earth, recycling conserves these resources.

6) Recycling metal cuts down on the amount of rubbish that gets sent to <u>landfill</u>. Landfill takes up space and <u>pollutes</u> the surroundings. If all the aluminium cans in the UK were recycled, there'd be 14 million fewer dustbins of waste each year.

For example...

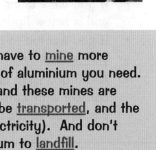

1) If you didn't recycle, say, <u>aluminium</u>, you'd have to <u>mine</u> more aluminium ore — <u>4 tonnes</u> for every <u>1 tonne</u> of aluminium you need. But mining makes a mess of the <u>landscape</u> (and these mines are often in <u>rainforests</u>). The ore then needs to be <u>transported</u>, and the aluminium <u>extracted</u> (which uses <u>loads</u> of electricity). And don't forget the cost of sending your <u>used</u> aluminium to <u>landfill</u>.

2) So it's a <u>complex</u> calculation, but for every 1 kg of aluminium cans you recycle, you <u>save</u>:

- <u>95%</u> or so of the <u>energy</u> needed to mine and extract 'fresh' aluminium,
- <u>4 kg</u> of aluminium ore,
- a <u>lot</u> of waste.

> In fact, aluminium's about the most cost-effective metal to recycle.

Recycling — do the Tour de France twice...

Recycling metal saves <u>natural resources</u> and <u>money</u> and reduces <u>environmental problems</u>. It's great.

Properties of Metals

Metals are all the same but slightly different. They have some basic properties in common, but each has its own specific combination of properties, which mean you use different ones for different purposes.

Metals are Strong and Bendy and They're Great Conductors

1) Most of the elements are metals — so they cover most of the periodic table.
   In fact, only the elements on the far right are non-metals.

2) All metals have some fairly similar basic properties:

   • Metals are strong (hard to break), but they can
     be bent or hammered into different shapes.

   • They're great at conducting heat.

   • They conduct electricity well.

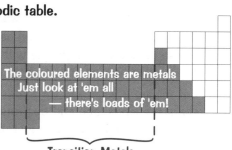

The coloured elements are metals
Just look at 'em all
— there's loads of 'em!

Transition Metals

3) Metals (and especially transition metals, which are found in the
   centre block of the periodic table) have loads of everyday uses because of these properties...

   • Their strength and 'bendability' makes them handy for making into things like bridges and car bodies.

   • Metals are ideal if you want to make something that heat needs to travel through,
     like a saucepan base.

   • And their conductivity makes them great for making things like electrical wires.

A Metal's Exact Properties Decide How It's Best Used

1) The properties above are typical properties of metals.
   Not all metals are the same though — here are three examples for you:

   Copper is a good conductor of electricity, so it's ideal for drawing out into electrical wires.
   It's hard and strong but can be bent. It also doesn't react with water.

Aluminium is corrosion-resistant and has a low density. Pure aluminium isn't particularly strong, but it forms hard, strong alloys (see page 27).	Titanium is another low density metal. Unlike aluminium it's very strong. It is also corrosion-resistant.

2) Different metals are chosen for different uses because of their specific properties. For example:
   • If you were doing some plumbing, you'd pick a metal that could be bent to make pipes and tanks,
     and is below hydrogen in the reactivity series so it doesn't react with water. Copper is great for this.
   • If you wanted to make an aeroplane, you'd probably use metal as it's strong and can be
     bent into shape. But you'd also need it to be light, so aluminium would be a good choice.
   • And if you were making replacement hips, you'd pick a metal that won't corrode when it comes in
     contact with water. It'd also have to be light too, and not too bendy. Titanium has all of these
     properties so it's used for this.

Metals are Good — but Not Perfect

1) Metals are very useful structural materials, but some corrode when exposed to air and water, so they
   need to be protected, e.g. by painting. If metals corrode, they lose their strength and hardness.

2) Metals can get 'tired' when stresses and strains are repeatedly put on them over time. This is known
   as metal fatigue and leads to metals breaking, which can be very dangerous, e.g. in planes.

Metal fatigue? — yeah, I've had enough of this page too...

So, all metals conduct electricity and heat and can be bent into shape. But lots of them have special properties
too. We have to decide what properties we need and use the metal with those properties.

Alloys

Pure metals often aren't quite right for certain jobs. Scientists don't just make do, oh no my friend... they <u>mix two metals together</u> (or mix a metal with a non-metal) — creating an <u>alloy</u> with the properties they want.

Pure Iron Tends to be a Bit Too Bendy

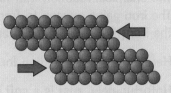

1) 'Iron' straight from the blast furnace is only <u>96% iron</u>. The other 4% is impurities such as <u>carbon</u>.

2) This impure iron is used as <u>cast iron</u>. It's handy for making <u>ornamental railings</u>, but it doesn't have many other uses because it's <u>brittle</u>.

3) So <u>all</u> the impurities are removed from most of the blast furnace iron. This pure iron has a <u>regular arrangement</u> of identical atoms. The layers of atoms can <u>slide over each other</u>, which makes the iron <u>soft</u> and <u>easily shaped</u>. This iron is far <u>too bendy</u> for most uses.

Most Iron is Converted into Steel — an Alloy

Most of the pure iron is changed into <u>alloys</u> called <u>steels</u>. Steels are formed by adding <u>small</u> amounts of <u>carbon</u> and sometimes <u>other metals</u> to the iron.

TYPE OF STEEL	PROPERTIES	USES
Low carbon steel (0.1% carbon)	easily shaped	car bodies (see page 28)
High carbon steel (1.5% carbon)	very hard, inflexible	blades for cutting tools, bridges
Stainless steel (chromium added, and sometimes nickel)	corrosion-resistant	cutlery, containers for corrosive substances

Alloys are Harder Than Pure Metals

1) Different elements have <u>different sized atoms</u>. So when an element such as carbon is added to pure iron, the <u>smaller</u> carbon atoms will <u>upset</u> the layers of pure iron atoms, making it more difficult for them to slide over each other. So alloys are <u>harder</u>.

2) Many metals in use today are actually <u>alloys</u>. E.g.:

BRONZE = COPPER + TIN Bronze is <u>harder</u> than copper. It's good for making medals and statues from.

CUPRONICKEL = COPPER + NICKEL This is <u>hard</u> and <u>corrosion resistant</u>. It's used to make "silver" coins.

GOLD ALLOYS ARE USED TO MAKE JEWELLERY Pure gold is <u>too soft</u>. Metals such as zinc, copper, silver, palladium and nickel are used to harden the "gold".

ALUMINIUM ALLOYS ARE USED TO MAKE AIRCRAFT Aluminium has a <u>low density</u>, but it's <u>alloyed</u> with small amounts of other metals to make it <u>stronger</u>.

3) In the past, the development of alloys was by <u>trial and error</u>. But nowadays we understand much more about the properties of metals, so alloys can be <u>designed</u> for specific uses.

A brass band — harder than Iron Maiden...

The <u>Eiffel Tower</u> is made of iron — but the problem with iron is, it goes <u>rusty</u> if air and water get to it. So the Eiffel Tower has to be <u>painted</u> every seven years to make sure that it doesn't rust. This is quite a job and takes an entire year for a team of 25 painters. Too bad they didn't use stainless steel.

Building Cars

There are <u>loads of different materials</u> in your average car — different materials have different <u>properties</u> and so have different <u>uses</u>. Makes sense.

Iron and Steel Corrode Much More than Aluminium

Iron corrodes easily. In other words, it <u>rusts</u>. ⇐ The word "rust" is only used for the corrosion of iron, not other metals.

Rusting only happens when the iron's in contact with both <u>oxygen</u> (from the air) and <u>water</u>.

The chemical reaction that takes place when iron corrodes is an <u>oxidation</u> reaction. The iron <u>gains oxygen</u> to form <u>iron(III) oxide</u>. Water then becomes loosely bonded to the iron(III) oxide and the result is <u>hydrated iron(III) oxide</u> — which we call rust.

Here's the <u>word equation</u> for the reaction: | iron + oxygen + water → hydrated iron(III) oxide |

Unfortunately, rust is a soft crumbly solid that soon <u>flakes off</u> to leave more iron available to <u>rust again</u>. And if the water's <u>salty</u> or <u>acidic</u>, rusting will take place a <u>lot quicker</u>. Cars in coastal places rust a lot because they get covered in <u>salty sea-spray</u>. Cars in <u>dry</u> deserty places hardly rust at all.

<u>Aluminium doesn't corrode</u> when it's wet. This is a bit odd because aluminium is more reactive than iron. What happens is that the aluminium reacts <u>very quickly</u> with <u>oxygen in the air</u> to form <u>aluminium oxide</u>. A nice <u>protective layer</u> of <u>aluminium oxide</u> sticks firmly to the aluminium below and <u>stops any further reaction taking place</u> (the oxide isn't crumbly and flaky like rust, so it won't fall off).

Car Bodies: Aluminium or Steel?

Aluminium has two big advantages over steel:

1) It has a much <u>lower density</u>, so the car body of an aluminium car will be <u>lighter</u> than the same car made of steel. This gives the aluminium car much better <u>fuel economy</u>, which <u>saves fuel resources</u>.

2) A car body made with aluminium <u>corrodes less</u> and so it'll have a <u>longer lifetime</u>.

But aluminium has a <u>massive disadvantage</u>. It <u>costs a lot more</u> than iron or steel. That's why car manufacturers tend to build cars out of steel instead.

You Need Various Materials to Build Different Bits of a Car

1) <u>Steel</u> is strong and it can be hammered into sheets and welded together — good for the <u>bodywork</u>.

2) <u>Aluminium</u> is <u>strong</u> and has a <u>low density</u> — it's used for <u>parts of the engine</u>, to reduce weight.

3) <u>Glass</u> is <u>transparent</u> — cars need <u>windscreens and windows</u>.

4) <u>Plastics</u> are <u>light and hardwearing</u>, so they're used as internal coverings for doors, dashboards, etc. They're also electrical insulators, used for covering electrical wires.

5) <u>Fibres</u> (natural and synthetic) are <u>hard-wearing</u>, so they're used to cover the <u>seats and floor</u>.

Unless you can afford leather seats, that is.

Recycling Cars is Important

1) As with all recycling, the idea is to <u>save natural resources</u>, <u>save money</u> and <u>reduce landfill use</u>.

2) At the moment a lot of the <u>metal</u> from a scrap car is recycled, though most of the other materials (e.g. plastics, rubber, etc.) go into <u>landfill</u>. But European laws are now in place saying that <u>85%</u> of the materials in a car (rising to <u>95%</u> of a car by 2015) must be recyclable.

3) The biggest <u>problem</u> with recycling all the non-metal bits of a car is that they have to be <u>separated</u> before they can be recycled. Sorting out different types of plastic is a pain in the neck.

CGP jokes — 85% recycled since 1996...

When manufacturers choose materials for cars, they have to weigh up alternatives — they balance <u>safety</u>, <u>environmental impact</u>, and <u>cost</u>. You never know, one day you could be asked to do the same. Sounds fun.

Revision Summary for Section Two

Okay, if you were just about to turn the page without doing these revision summary questions, then stop. What kind of attitude is that... Is that really the way you want to live your life... running, playing and having fun... Of course not. That's right. Do the questions. It's for the best all round.

1) Name a material we use that comes from:
   a) plants   b) animals

2) What is the advantage of synthetic rubber over natural rubber?

3) Write down the symbol equation showing the thermal decomposition of limestone.

4) What products are produced when limestone reacts with an acid?

5) What is calcium hydroxide used for?

6) Name three building materials made from limestone.

7) Plans to develop a limestone quarry and a cement factory on some hills next to your town are announced. Describe the views that the following might have:
   a) dog owners   b) a mother of young children
   c) the owner of a cafe   d) a beetle

8) What is rock salt?

9) Give three methods of obtaining salt industrially.

10) What's the definition of an ore?

11) Explain why zinc can be extracted by reduction with carbon but magnesium can't.

12) Give a reason why aluminium is an expensive metal.

13) What is electrolysis?

14) Describe the process of purifying copper by electrolysis.

15) Describe how scrap iron is used to displace copper from solution.

16) What is the name of the method where plants are used to extract metals from soil?

17) Give three reasons why it's good to recycle metal.

18) Give three properties of metals.

19) Briefly describe two problems with metals.

20) What is the problem with using a) iron straight from the blast furnace,   b) very pure iron?

21) Give two examples of alloys and say what's in them.

22) Write down the word equation for the corrosion of iron.

23) Explain why a car parked on the Brighton seafront rusts more than a car parked in hot, dry Cairo.

24) Why doesn't aluminium corrode when it's wet?

25) Give two advantages of using aluminium instead of steel for car bodywork.

Fractional Distillation of Crude Oil

Crude oil is formed from the buried remains of plants and animals — it's a <u>fossil fuel</u>. Over millions of years, the remains turn to crude oil, which can be extracted by drilling and pumping.

Crude Oil is a Mixture of Hydrocarbons

1) A mixture consists of <u>two</u> (or more) elements or compounds that <u>aren't chemically bonded</u> to each other.

2) <u>Crude oil</u> is a <u>mixture</u> of many different compounds. <u>Most</u> of the compounds are <u>hydrocarbon</u> molecules.

3) <u>Hydrocarbons</u> are basically <u>fuels</u> such as petrol and diesel. They're made of just <u>carbon</u> and <u>hydrogen</u>.

4) There are <u>no chemical bonds</u> between the different parts of a mixture, so the different hydrocarbon molecules in crude oil <u>aren't</u> chemically bonded to one another.

5) This means that they all keep their <u>original properties</u>, such as their condensing points. The <u>properties</u> of a mixture are just a <u>mixture</u> of the properties of the <u>separate parts</u>.

6) The parts of a mixture can be <u>separated</u> out by <u>physical methods</u>, e.g. crude oil can be split up into its separate <u>fractions</u> by <u>fractional distillation</u>. Each fraction contains molecules with a <u>similar number of carbon atoms</u> to each other (see next page).

Crude Oil is Split into Separate Groups of Hydrocarbons

The <u>fractionating column</u> works <u>continuously</u>, with heated crude oil piped in at the <u>bottom</u>. The vaporised oil rises up the column and the various <u>fractions</u> are <u>constantly tapped off</u> at the different levels where they <u>condense</u>.

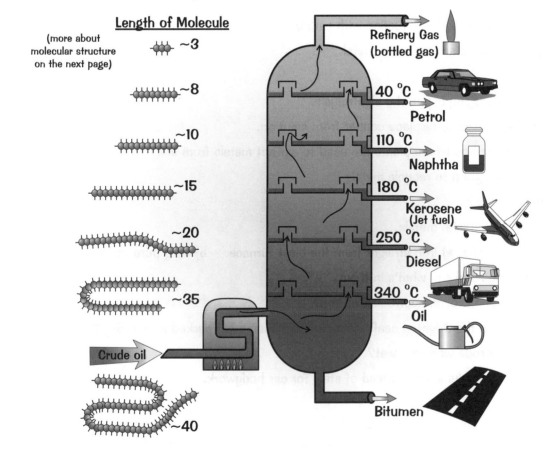

Length of Molecule

(more about molecular structure on the next page)

~3

~8

~10

~15

~20

~35

Crude oil

~40

Refinery Gas (bottled gas)

40 °C — Petrol

110 °C — Naphtha

180 °C — Kerosene (Jet fuel)

250 °C — Diesel

340 °C — Oil

Bitumen

Crude oil — it's always cracking dirty jokes...

It's amazing what you get from buried dead stuff. But it has had a <u>few hundred million years</u> with high <u>temperature</u> and <u>pressure</u> to get into the useful state it's in now. So if we use it all, we're going to have to wait an awful <u>long time</u> for more to form. <u>No one knows</u> exactly when oil will run out, but some scientists reckon that it could be within <u>this century</u>. The thing is, <u>technology</u> is advancing all the time, so one day it's likely that we'll be able to extract oil that's too difficult and expensive to extract at the moment.

Properties and Uses of Crude Oil

The <u>different fractions</u> of crude oil have <u>different properties</u>. Handy, eh?

Hydrocarbon Properties Change as the Chain Gets Longer

As the <u>length</u> of the carbon chain changes, the <u>properties</u> of the hydrocarbon change.

1) The <u>shorter</u> the molecules, the <u>more runny</u> the hydrocarbon is — that is, the <u>less viscous</u> (gloopy) it is.

2) The <u>shorter</u> the molecules, the <u>more volatile</u> they are. "More volatile" means they turn into a gas at a <u>lower temperature</u>. So, the shorter the molecules, the <u>lower</u> the <u>temperature</u> at which that fraction <u>vaporises</u> or <u>condenses</u> — and the <u>lower</u> its <u>boiling point</u>.

3) Also, the <u>shorter</u> the molecules, the more <u>flammable</u> (easier to ignite) the hydrocarbon is.

The Properties Depend on the Forces Between Molecules

It's all down to the forces in between hydrocarbons...

1) There are two important types of <u>bond</u> in crude oil:
   a) The <u>strong covalent bonds</u> between the carbons and hydrogens <u>within each hydrocarbon</u> molecule.
   b) The <u>intermolecular forces</u> of attraction between <u>different hydrocarbon molecules</u> in the mixture.

2) When the crude oil mixture is <u>heated</u>, the molecules are supplied with <u>extra energy</u>.

3) This makes the molecules <u>move about</u> more. Eventually a molecule might have enough energy to <u>overcome</u> the <u>intermolecular forces</u> that keep it with the other molecules.

4) It can now go <u>whizzing off</u> as a <u>gas</u>.

5) The <u>covalent bonds</u> holding each molecule together are <u>much stronger</u> than the intermolecular forces, so they <u>don't</u> break. That's why you don't end up with lots of <u>little molecules</u>.

6) The intermolecular forces break a lot more <u>easily</u> in <u>small</u> molecules than they do in bigger molecules. That's because the intermolecular forces of attraction are much <u>stronger</u> between big molecules than they are between small molecules.

7) It makes sense if you think about it — even if a big molecule can overcome the forces attracting it to another molecule at a <u>few points</u> along its length, it's still got lots of <u>other</u> places where the force is still strong enough to hold it in place.

8) That's why <u>big</u> molecules have <u>higher boiling points</u> than small molecules do — more energy is needed for them to break out of a liquid and form a gas.

not many intermolecular forces to break

lots of intermolecular forces to break

The Uses Of Hydrocarbons Depend on their Properties

1) The <u>volatility</u> helps decide what the fraction is used for. The <u>refinery gas fraction</u> has the shortest molecules, so it has the <u>lowest boiling point</u> — in fact it's a gas at room temperature. This makes it ideal for using as <u>bottled gas</u>. It's stored under pressure as liquid in 'bottles'. When the tap on the bottle is opened, the fuel vaporises and flows to the burner where it's ignited.

2) The <u>petrol</u> fraction has longer molecules, so it has a higher boiling point. Petrol is a <u>liquid</u> which is ideal for storing in the fuel tank of a car. It can flow to the engine where it's easily <u>vaporised</u> to mix with the air before it is ignited.

3) The <u>viscosity</u> also helps decide how the hydrocarbons are <u>used</u>. The really gloopy, viscous hydrocarbons are used for <u>lubricating engine parts</u> and for <u>covering roads</u>.

Positively boiling over with chemistry fun...

There's loads of info for you on this page. The trickiest thing is getting your head around how the <u>boiling point</u> is affected by <u>intermolecular forces</u> and the <u>size</u> of hydrocarbon chains. But it's not actually all that complicated. Read it a few times, scribble it down a few more and it should start to make sense. Promise.

Using Crude Oil as a Fuel

Nothing as amazingly useful as crude oil would be without its problems. No, that'd be too good to be true.

Crude Oil Provides an Important Fuel for Modern Life

1) Crude oil fractions burn cleanly so they make good <u>fuels</u>. Most modern transport is fuelled by a crude oil fraction, e.g. cars, boats, trains and planes. Parts of crude oil are also burned in <u>central heating systems</u> in homes and in <u>power stations</u> to <u>generate electricity</u>.

2) There's a <u>massive industry</u> with scientists working to find oil reserves, take it out of the ground, and turn it into useful products. As well as fuels, crude oil also provides the raw materials for making various <u>chemicals</u>, including <u>plastics</u>.

3) Often, <u>alternatives</u> to using crude oil fractions as fuel are possible. E.g. electricity can be generated by <u>nuclear</u> power or <u>wind</u> power, there are <u>ethanol</u>-powered cars, and <u>solar</u> energy can be used to heat water.

4) But things tend to be <u>set up</u> for using oil fractions. For example, cars are designed for <u>petrol or diesel</u> and it's <u>readily available</u>. There are filling stations all over the country, with storage facilities and pumps specifically designed for these crude oil fractions. So crude oil fractions are often the <u>easiest and cheapest</u> thing to use.

5) Crude oil fractions are often <u>more reliable</u> too — e.g. solar and wind power won't work without the right weather conditions. Nuclear energy is reliable, but there are lots of concerns about its <u>safety</u> and the storage of radioactive waste.

But it Might Run Out One Day... Eeek

1) Most scientists think that oil will <u>run out</u> — it's a <u>non-renewable fuel</u>.

2) No one knows exactly when it'll run out but there have been heaps of <u>different predictions</u> — e.g. about 40 years ago, scientists predicted that it'd all be gone by the year 2000.

3) <u>New oil reserves</u> are discovered from time to time and <u>technology</u> is constantly improving, so it's now possible to extract oil that was once too <u>difficult</u> or <u>expensive</u> to extract.

4) In the <u>worst-case scenario</u>, oil may be pretty much gone in about 25 years — and that's not far off.

5) Some people think we should <u>immediately stop</u> using oil for things like transport, for which there are alternatives, and keep it for things that it's absolutely <u>essential</u> for, like some chemicals and medicines.

6) It will take time to <u>develop</u> alternative fuels that will satisfy all our energy needs (see page 53 for more info). It'll also take time to <u>adapt things</u> so that the fuels can be used on a wide scale. E.g. we might need different kinds of car engines, or special storage tanks built.

7) One alternative is to generate energy from <u>renewable</u> sources — these are sources that <u>won't run out</u>. Examples of renewable energy sources are <u>wind power</u>, <u>solar power</u> and <u>tidal power</u>.

8) So however long oil does last for, it's a good idea to start <u>conserving</u> it and finding <u>alternatives</u> now.

Crude Oil is NOT the Environment's Best Friend

1) <u>Oil spills</u> can happen as the oil is being transported by tanker — this spells <u>disaster</u> for the local environment. <u>Birds</u> get covered in the stuff and are <u>poisoned</u> as they try to clean themselves. Other creatures, like <u>sea otters</u> and <u>whales</u>, are poisoned too.

2) You have to <u>burn oil</u> to release the energy from it. But burning oil is thought to be a major cause of <u>global warming</u>, <u>acid rain</u> and <u>global dimming</u> — see pages 51 and 53.

If oil alternatives aren't developed, we might get caught short...

Crude oil is <u>really important</u> to our lives. Take <u>petrol</u> for instance — at the first whisper of a shortage, there's mayhem. Loads of people dash to the petrol station and start filling up their tanks. This causes a queue, which starts everyone else panicking. I don't know what they'll do when it runs out totally.

Cracking Crude Oil

After the distillation of crude oil (see page 30), you've still got both short and long hydrocarbons, just not all mixed together. But there's <u>more demand</u> for some products, like <u>petrol</u>, than for others.

Cracking Means Splitting Up Long–chain Hydrocarbons...

1) <u>Long-chain hydrocarbons</u> form <u>thick gloopy liquids</u> like <u>tar</u> which aren't all that useful, so...

2) ... a lot of the longer molecules produced from <u>fractional distillation</u> are <u>turned into smaller ones</u> by a process called <u>cracking</u>.

3) Some of the products of cracking are useful as fuels, e.g. petrol for cars and paraffin for jet fuel.

4) Cracking also produces substances like <u>ethene</u>, which are needed for <u>making plastics</u> (see page 35).

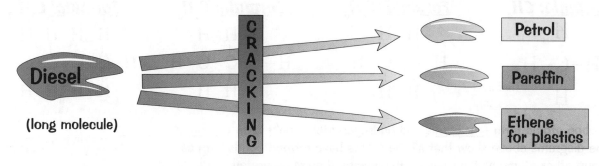

Diesel (long molecule) → CRACKING → Petrol, Paraffin, Ethene for plastics

...by Passing Vapour Over a Hot Catalyst

1) <u>Cracking</u> is a <u>thermal decomposition</u> reaction — <u>breaking molecules down</u> by <u>heating</u> them.

2) The first step is to <u>heat</u> the long-chain hydrocarbon to <u>vaporise</u> it (turn it into a gas).

3) Then the <u>vapour</u> is passed over a <u>powdered catalyst</u> at a temperature of about <u>400 °C – 700 °C</u>.

4) <u>Aluminium oxide</u> is the catalyst used.

5) The <u>long-chain</u> molecules <u>split apart</u> or "crack" on the <u>surface</u> of the specks of catalyst.

Vaporised kerosene → Aluminium oxide → Octane + Ethene

6) Most of the <u>products</u> of cracking are <u>alkanes</u> (see page 34) and unsaturated hydrocarbons called <u>alkenes</u> (see page 34)...

An alternative way of cracking long-chain hydrocarbons is to mix the vapour with steam at a very high temperature.

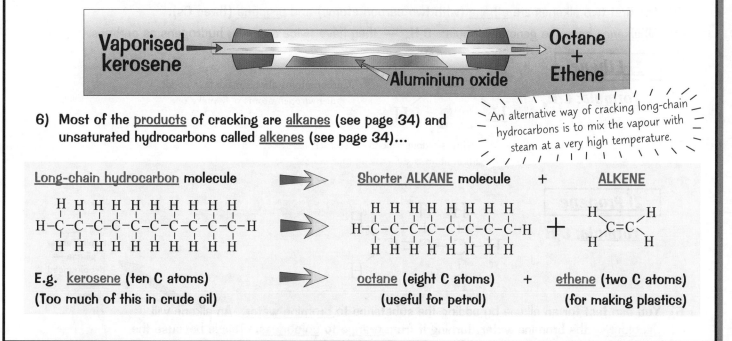

| Long-chain hydrocarbon molecule | → | Shorter ALKANE molecule + ALKENE |

E.g. <u>kerosene</u> (ten C atoms)   →   octane (eight C atoms)   +   ethene (two C atoms)
(Too much of this in crude oil)   (useful for petrol)   (for making plastics)

Get cracking — there's a lot to learn...

Crude oil is <u>useful stuff</u>, there's no doubt about it. But using it is not without its problems (see page 32 for more about fuels). For example, oil is shipped around the planet, which can lead to <u>slicks</u> if there's an accident. Also, burning oil is thought to cause <u>climate change</u>, <u>acid rain</u> and <u>global dimming</u>. Oil is going to start <u>running out</u> one day, which will lead to big difficulties.

Alkanes and Alkenes

Alkanes and alkenes might sound the same, but they're not. This page has the lowdown.

Crude Oil *is Mostly* Alkanes

1) All the fractions of crude oil are hydrocarbons called alkanes.

2) Alkanes are made up of chains of carbon atoms surrounded by hydrogen atoms. ⟹

3) Different alkanes have chains of different lengths.

4) The first four alkanes are methane (natural gas), ethane, propane and butane.

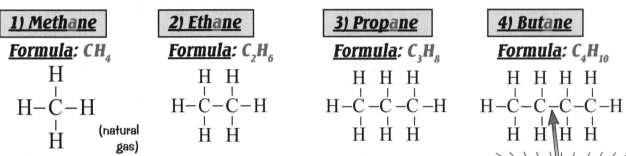

1) Methane
Formula: CH_4

(natural gas)

2) Ethane
Formula: C_2H_6

3) Propane
Formula: C_3H_8

4) Butane
Formula: C_4H_{10}

Each straight line shows a covalent bond (page 13).

5) Carbon atoms form four bonds and hydrogen atoms only form one bond. The diagrams above show that all the atoms have formed bonds with as many other atoms as they can — this means they're saturated.

6) Alkanes all have the general formula C_nH_{2n+2}. So if an alkane has 5 carbons, it's got to have $(2 \times 5) + 2 = 12$ hydrogens.

Alkenes *Have a* C=C Double Bond

1) Alkenes are hydrocarbons which have a double bond between two of the carbon atoms in their chain.

2) They are known as unsaturated because they can make more bonds — the double bond can open up, allowing the two carbon atoms to bond with other atoms.

3) The first two alkenes are ethene (with two carbon atoms) and propene (three Cs).

4) All alkenes have the general formula: C_nH_{2n} — they have twice as many hydrogens as carbons.

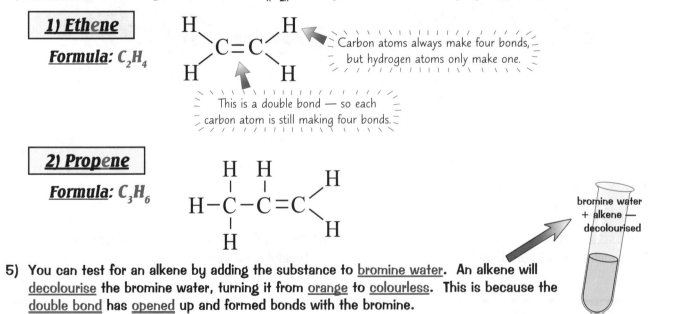

1) Ethene
Formula: C_2H_4

Carbon atoms always make four bonds, but hydrogen atoms only make one.

This is a double bond — so each carbon atom is still making four bonds.

2) Propene
Formula: C_3H_6

bromine water + alkene — decolourised

5) You can test for an alkene by adding the substance to bromine water. An alkene will decolourise the bromine water, turning it from orange to colourless. This is because the double bond has opened up and formed bonds with the bromine.

Alkane ya if you don't learn this...

Don't get alkenes confused with alkanes — that one letter makes all the difference. Alkenes have a C=C bond, alkanes don't. The first parts of their names are the same though — that just tells you the number of C atoms.

Using Alkenes to Make Polymers

Before we knew how to make <u>polymers</u>, there were no <u>polythene bags</u>.

Alkenes *Can Be Used to Make* Polymers

1) Probably the most useful thing you can do with alkenes is <u>polymerisation</u>. This means joining together lots of <u>small alkene molecules</u> (<u>monomers</u>) to form <u>very large molecules</u> — these long-chain molecules are called <u>polymers</u>.

> Polymers are often written without the brackets — e.g. polyethene.

2) For instance, many <u>ethene</u> molecules can be joined up to produce <u>poly(ethene)</u> or "polythene".

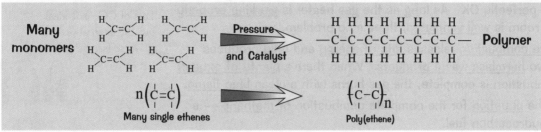

Many monomers → Pressure and Catalyst → Polymer

3) In the same way, if you join lots of <u>propene</u> molecules together, you've got <u>poly(propene)</u>.

Different Polymers *Have Different* Physical Properties...

1) The physical properties of a polymer depend on <u>what it's made from</u>. Polyamides are usually stronger than poly(ethene), for example.

2) A polymer's <u>physical properties</u> are also affected by the <u>temperature and pressure</u> of polymerisation. Poly(ethene) made at <u>200 °C</u> and <u>2000 atmospheres pressure</u> is <u>flexible</u>, and has <u>low density</u>. But poly(ethene) made at <u>60 °C</u> and a <u>few atmospheres pressure</u> with a <u>catalyst</u> is <u>rigid</u> and <u>dense</u>.

...Which Make Them Suitable for Various Different Uses

1) <u>Light, stretchable</u> polymers such as low density poly(ethene) are used to make plastic bags. <u>Elastic</u> polymer fibres are used to make super-stretchy <u>LYCRA® fibre</u> for tights.

2) <u>PVC</u> is <u>strong</u> and durable, and it can be made either <u>rigid</u> or <u>stretchy</u>. The rigid kind is used to make <u>window frames and piping</u>. The stretchy kind is used to make <u>synthetic leather</u>.

3) <u>Polystyrene foam</u> is used in <u>packaging</u> to protect breakable things, and it's used to make disposable coffee cups (the trapped air in the foam makes it a brilliant <u>thermal insulator</u>).

4) <u>New uses</u> are developed all the time. <u>Waterproof</u> coatings for fabrics are made of polymers. <u>Dental polymers</u> are used in resin <u>tooth fillings</u>. Polymer <u>hydrogel wound dressings</u> keep wounds moist.

5) <u>New biodegradable packaging</u> materials made from polymers and <u>cornstarch</u> are being produced.

6) <u>Memory foam</u> is an example of a <u>smart material</u>. It's a polymer that gets <u>softer</u> as it gets <u>warmer</u>. Mattresses can be made of memory foam — they mould to your body shape when you lie on them.

Polymers *Are Cheap, but Most* Don't Rot — *They're Hard to Get Rid Of*

1) Most polymers aren't "<u>biodegradable</u>" — they're not broken down by microorganisms, so they <u>don't rot</u>.

2) It's difficult to get rid of them — if you bury them in a landfill site, they'll <u>still</u> be there <u>years later</u>. The best thing is to <u>re-use</u> them as many times as possible and then <u>recycle</u> them if you can.

3) Things made from polymers are usually <u>cheaper</u> than things made from metal. However, as <u>crude oil resources</u> get <u>used up</u>, the <u>price</u> of crude oil will rise. Crude oil products like polymers will get dearer.

4) It may be that one day there won't be <u>enough</u> oil for fuel AND plastics AND all the other uses. Choosing how to use the oil that's left means weighing up advantages and disadvantages on all sides.

Revision's like a polymer — you join lots of little facts up...

Polymers are <u>all over the place</u>. You've even got polymers <u>on the inside</u> — DNA's a polymer.

Burning Fuels

We get loads of fuels from oil. And then we burn them. But there's <u>burning</u> and there's <u>burning</u>...

Complete Combustion Happens When There's Plenty of Oxygen

1) When there's <u>plenty of oxygen</u> about, hydrocarbons burn to produce only <u>carbon dioxide</u> and <u>water</u>.

   hydrocarbon + oxygen ⟹ carbon dioxide + water   (+ energy)

2) The <u>hydrogen</u> and <u>carbon</u> in the hydrocarbon have both been <u>oxidised</u>.

3) Many <u>gas room heaters</u> release these <u>waste gases</u> into the room, which is perfectly OK. As long as the gas heater is <u>working properly</u> and the room is <u>well ventilated</u>, there's no problem.

4) <u>Complete combustion</u> releases <u>lots of energy</u> and only produces those two <u>harmless waste products</u>. When there's <u>plenty of oxygen</u> and combustion is complete, the gas burns with a <u>clean blue flame</u>.

5) Here's the <u>equation</u> for the complete combustion of <u>methane</u> — a simple hydrocarbon fuel.

Lots of CO_2 isn't ideal, but the alternatives are worse (see below).

Natural gas is mostly methane (CH_4).

$$CH_4 + 2O_2 \rightarrow 2H_2O + CO_2$$

You can test for CO_2 by bubbling the gas through limewater. It turns limewater milky (see page 19).

Partial Combustion of Hydrocarbons is NOT Safe

Partial combustion is also known as incomplete combustion.

1) If there <u>isn't enough oxygen</u>, partial combustion will occur. Carbon dioxide and water are still produced, but you can also get <u>carbon monoxide</u> (CO) and <u>carbon</u>.

2) Partial combustion means a <u>smoky yellow flame</u>, and <u>less energy</u> than complete combustion.

   hydrocarbon + oxygen ⟹ carbon + carbon monoxide + carbon dioxide + water

   (+ energy)

3) The <u>carbon monoxide</u> is a <u>colourless</u>, <u>odourless</u> and very toxic (<u>poisonous</u>) gas.

4) Every year people are <u>killed</u> while they sleep due to <u>faulty</u> gas fires and boilers filling the room with <u>carbon monoxide</u> and nobody realising — this is why it's important to <u>regularly service gas appliances</u>. The black carbon given off produces <u>sooty marks</u> — a <u>clue</u> that the fuel is <u>not</u> burning fully.

5) So basically, you want <u>lots of oxygen</u> when you're burning fuel — you get <u>more energy</u> given out, and you don't get any <u>messy soot</u> or <u>poisonous gases</u>.

6) Here's an example of an <u>equation</u> for partial combustion too.

$$4CH_4 + 6O_2 \rightarrow C + 2CO + CO_2 + 8H_2O$$

This is just <u>one possibility</u>. The products depend on how much oxygen is present...

... e.g. you could also have: $4CH_4 + 7O_2 \rightarrow 2CO + 2CO_2 + 8H_2O$ — the important thing is that the equation is <u>balanced</u> (see p. 14).

There's Lots to Consider When Choosing the Best Fuel

1) <u>Ease of ignition</u> — whether it <u>burns easily</u>. Fuels like gas burn more easily than diesel.

2) <u>Energy value</u> (i.e. amount of energy released) — see page 96 for more on this.

3) <u>Ash and smoke</u> — some fuels, like coal, leave behind a lot of <u>ash</u> that needs to be disposed of.

4) <u>Storage and transport</u> — gas needs to be stored in special <u>canisters</u> and coal needs to be kept <u>dry</u>. Fuels need to be transported <u>carefully</u> as gas leaks and oil spills can be dangerous.

Blue flame good, orange flame bad...

An ideal fuel would be <u>easy to ignite</u>, produce <u>no soot</u> or <u>toxic products</u>, release <u>loads of energy</u>, leave no <u>residue</u> (what's left after burning), and be capable of being <u>stored safely</u>. There should also be a <u>cheap</u> and <u>sustainable supply</u> of it. In reality, you have to go for the <u>best compromise</u>.

Revision Summary for Section Three

Now, my spies tell me that some naughty people skip these pages without so much as reading through the list of questions. Well, you shouldn't, because what's the point in reading the whole section if you're not going to check if you really know it or not? Look, just read the first 10 questions, and I guarantee there'll be an answer you'll have to look up. And when it comes up in the exam, you'll be so glad you did. Plus, if you don't do as you're told my spies will tell me, and then you won't get any toys. Or something.

1) What does crude oil consist of? What does fractional distillation do to crude oil?

2) Is a short-chain hydrocarbon more viscous than a long-chain hydrocarbon? Is it more volatile?

3)* You're going on holiday to a very cold place. The temperature will be about –10 °C. Which of the fuels shown on the right do you think will work best in your camping stove? Explain your answer.

Fuel	Boiling point (°C)
Propane	–42
Butane	–0.4
Pentane	36.2

4) Name the two important types of bond in crude oil.

5) Why do big hydrocarbon molecules have higher boiling points than small ones?

6) What is "cracking"? Why is it done?

7) Give a typical example of a substance that is cracked, and the products that you get from cracking it.

8) What's the general formula for an alkane?

9) What kind of carbon-carbon bond do alkenes have?

10) What is the general formula for alkenes?

11) Draw the displayed formula of ethene.

12) What are polymers? What kinds of substances can form polymers?

13) Give two factors which affect the physical properties of a polymer.

14) List four uses of polymers.

15) Why might polymers become more expensive in the future?

16) What is produced during complete combustion of hydrocarbons?

17) Explain how partial combustion can be harmful to humans.

18) Give four things you might want to consider when deciding on the best fuel to use.

* Answers on page 140.

Plant Oils

If you squeeze a walnut really hard, out will ooze some walnut oil, which you could use to make walnut mayonnaise. Much better to just buy some oil from the shop though.

We Can Extract Oils from Plants

olive mush

weight

olive oil

1) Some fruits and seeds contain a lot of oil. For example, avocados and olives are oily fruits. Brazil nuts, peanuts and sesame seeds are oily seeds (a nut is just a big seed really).

2) These oils can be extracted and used for food or for fuel.

3) To get the oil out, the plant material is crushed. The next step is to press the crushed plant material between metal plates and squash the oil out. This is the traditional method of producing olive oil.

4) Oil can be separated from crushed plant material by a centrifuge — rather like using a spin-dryer to get water out of wet clothes.

5) Or solvents can be used to get oil from plant material.

6) Distillation refines oil, and removes water, solvents and impurities.

Vegetable Oils Are Used in Food

1) Vegetable oils provide a lot of energy — they have a very high energy content.

2) There are other nutrients in vegetable oils. For example, oils from seeds contain vitamin E.

3) Vegetable oils contain essential fatty acids, which the body needs for many metabolic processes.

Vegetable Oils Have Benefits for Cooking

1) Vegetable oils have higher boiling points than water. This means they can be used to cook foods at higher temperatures and at faster speeds.

2) Cooking with vegetable oil gives food a different flavour. This is because of the oil's own flavour, but it's also down to the fact that many flavours come from chemicals that are soluble in oil. This means the oil 'carries' the flavour, making it seem more intense.

3) Using oil to cook food increases the energy we get from eating it.

Vegetable Oils Can Be Used to Produce Fuels

1) Vegetable oils such as rapeseed oil and soybean oil can be processed and turned into fuels.

2) Because vegetable oils provide a lot of energy they're really suitable for use as fuels.

3) A particularly useful fuel made from vegetable oils is called biodiesel. Biodiesel has similar properties to ordinary diesel fuel — it burns in the same way, so you can use it to fuel a diesel engine.

See page 53 for more about biodiesel.

That lippie fried a few sausages back in her heyday...

Plant oils have loads of different uses, from frying bacon to fuelling cars. Even waste oil, left over from manufacturing and cooking in fast food restaurants, ends up being used in pet food and cosmetics. Grim.

Plant Oils

Oils are usually quite runny at room temperature. That's fine for salad dressing, say, but not so good for spreading in your sandwiches. For that, you could <u>hydrogenate</u> the oil to make <u>margarine</u>...

Unsaturated Oils Contain C=C Double Bonds

1) Oils and fats contain <u>long-chain molecules</u> with lots of <u>carbon</u> atoms.

2) Oils and fats are either <u>saturated</u> or <u>unsaturated</u>.

3) Unsaturated oils contain <u>double bonds</u> between some of the carbon atoms in their carbon chains.

4) So, an unsaturated oil will <u>decolourise</u> bromine water (as the bromine opens up the double bond and joins on).

5) <u>Monounsaturated</u> fats contain <u>one</u> C=C double bond somewhere in their carbon chains. <u>Polyunsaturated</u> fats contain <u>more than one</u> C=C double bond.

bromine water + unsaturated oil — decolourised

Unsaturated Oils Can Be Hydrogenated

1) <u>Unsaturated</u> vegetable oils are <u>liquid</u> at room temperature.

2) They can be hardened by reacting them with <u>hydrogen</u> in the presence of a <u>nickel catalyst</u> at about <u>60 °C</u>. This is called <u>hydrogenation</u>. The hydrogen reacts with the double-bonded carbons and opens out the double bonds.

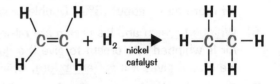

3) Hydrogenated oils have <u>higher melting points</u> than unsaturated oils, so they're <u>more solid</u> at room temperature. This makes them useful as <u>spreads</u> and for baking cakes and pastries.

4) Margarine is usually made from <u>partially</u> hydrogenated vegetable oil — turning <u>all</u> the double bonds in vegetable oil to single bonds would make margarine <u>too hard</u> and difficult to spread. Hydrogenating <u>most</u> of them gives margarine a nice, buttery, spreadable consistency.

5) Partially hydrogenated vegetable oils are often used instead of butter in processed foods, e.g. biscuits. These oils are a lot <u>cheaper</u> than butter and they <u>keep longer</u>. This makes biscuits cheaper and gives them a long shelf life.

6) But partially hydrogenating vegetable oils means you end up with a lot of so-called <u>trans fats</u>. And there's evidence to suggest that trans fats are <u>very bad</u> for you.

Vegetable Oils in Foods Can Affect Health

1) Vegetable oils tend to be <u>unsaturated</u>, while animal fats tend to be <u>saturated</u>.

2) In general, <u>saturated fats</u> are less healthy than <u>unsaturated fats</u> (as <u>saturated</u> fats <u>increase</u> the amount of <u>cholesterol</u> in the blood, which can block up the arteries and increase the risk of <u>heart disease</u>).

3) Natural <u>unsaturated</u> fats such as olive oil and sunflower oil <u>reduce</u> the amount of blood cholesterol. But because of the trans fats, <u>partially hydrogenated vegetable oil</u> increases the amount of <u>cholesterol</u> in the blood. So eating a lot of foods made with partially hydrogenated vegetable oils can actually increase the risk of heart disease.

4) <u>Cooking</u> food in oil, whether saturated, unsaturated or partially hydrogenated, makes it more <u>fattening</u>.

Double bonds — licensed to saturate...

This is tricky stuff. In a nutshell... there are saturated and unsaturated fats, which are <u>generally</u> bad and good for you (in that order) — easy enough. But... <u>partially hydrogenated vegetable oil</u> (which is unsaturated) is bad for you. Too much of the wrong types of fats can lead to heart disease. Got that...

Emulsions

Emulsions are all over the place in foods, cosmetics and paint. Don't say I didn't warn you...

Emulsions Can Be Made from Oil and Water

1) Oils don't dissolve in water. So far so good...

2) However, you can mix an oil with water to make an emulsion. Emulsions are made up of lots of droplets of one liquid suspended in another liquid. You can have an oil-in-water emulsion (oil droplets suspended in water) or a water-in-oil emulsion (water droplets suspended in oil).

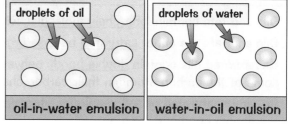

3) Emulsions are thicker than either oil or water. E.g. mayonnaise is an emulsion of sunflower oil (or olive oil) and vinegar — it's thicker than either.

4) The physical properties of emulsions make them suited to lots of uses in food — e.g. as salad dressings and in sauces. For instance, a salad dressing made by shaking olive oil and vinegar together forms an emulsion that coats salad better than plain oil or plain vinegar.

5) Generally, the more oil you've got in an oil-in-water emulsion, the thicker it is. Milk is an oil-in-water emulsion with not much oil and a lot of water — there's about 3% oil in full-fat milk. Single cream has a bit more oil — about 18%. Double cream has lots of oil — nearly 50%.

6) Whipped cream and ice cream are oil-in-water emulsions with an extra ingredient — air. Air is whipped into cream to give it a fluffy, frothy consistency for use as a topping. Whipping air into ice cream gives it a softer texture, which makes it easier to scoop out of the tub.

7) Emulsions also have non-food uses. Most moisturising lotions are oil-in-water emulsions. The smooth texture of an emulsion makes it easy to rub into the skin.

Some Foods Contain Emulsifiers to Help Oil and Water Mix

Oil and water mixtures naturally separate out. But here's where emulsifiers come in...

1) Emulsifiers are molecules with one part that's attracted to water and another part that's attracted to oil or fat. The bit that's attracted to water is called hydrophilic, and the bit that's attracted to oil is called hydrophobic.

2) The hydrophilic end of each emulsifier molecule latches onto water molecules.

3) The hydrophobic end of each emulsifier molecule cosies up to oil molecules.

4) When you shake oil and water together with a bit of emulsifier, the oil forms droplets, surrounded by a coating of emulsifier... with the hydrophilic bit facing outwards. Other oil droplets are repelled by the hydrophilic bit of the emulsifier, while water molecules latch on. So the emulsion won't separate out. Clever.

Using Emulsifiers Has Pros and Cons

1) Emulsifiers stop emulsions from separating out and this gives them a longer shelf-life.

2) Emulsifiers allow food companies to produce food that's lower in fat but that still has a good texture.

3) The down side is that some people are allergic to certain emulsifiers. For example, egg yolk is often used as an emulsifier — so people who are allergic to eggs need to check the ingredients very carefully.

Emulsion paint — spread mayonnaise all over the walls...

Before fancy stuff from abroad like olive oil, we fried our bacon and eggs in lard. Mmmm. Lard wouldn't be so good for making salad cream though. Emulsions like salad cream have to be made from shaking up two liquids — tiny droplets of one liquid are 'suspended' (NOT dissolved) in the other liquid.

Alcohols

This page is about different types of <u>alcohols</u> — and that's not just beer, wine and spirits.

Alcohols *Have an '-OH' Functional Group and End in '-ol'*

1) The <u>general formula</u> of an alcohol is $C_nH_{2n+1}OH$. So an alcohol with 2 carbons has the formula C_2H_5OH.

2) All alcohols contain the same <u>-OH group</u>. Here are the <u>first 3</u> in the homologous series:

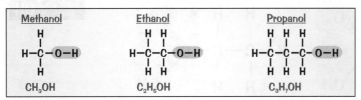

A homologous series is a group of chemicals that react in a similar way because they have the same functional group (in alcohols it's the −OH group).

3) The basic <u>naming</u> system is the same as for alkanes — but replace the final '-<u>e</u>' with '-<u>ol</u>'.

4) Don't write CH_4O instead of CH_3OH — it doesn't show the <u>functional -OH group</u>.

The First Three Alcohols Have Similar Properties

1) Alcohols are <u>flammable</u>. They burn in air to produce <u>carbon dioxide</u> and <u>water</u>.
E.g.
$$2CH_3OH_{(l)} + 3O_{2(g)} \rightarrow 2CO_{2(g)} + 4H_2O_{(g)}$$

2) The first three alcohols all <u>dissolve completely in water</u> to form <u>neutral solutions</u>.

3) They also react with <u>sodium</u> to give <u>hydrogen</u> and <u>alkoxides</u>, e.g. ethanol gives sodium ethoxide and H_2.
E.g.
$$2C_2H_5OH_{(l)} + 2Na_{(s)} \rightarrow 2C_2H_5ONa_{(aq)} + H_{2(g)}$$

4) <u>Ethanol</u> is the main alcohol in alcoholic drinks. It's not as <u>toxic</u> as methanol (which causes <u>blindness</u> if drunk) but it still damages the <u>liver</u> and <u>brain</u>.

Alcohols *are Used as Solvents*

1) Alcohols such as methanol and ethanol can <u>dissolve</u> most compounds that <u>water</u> dissolves, but they can also dissolve substances that <u>water can't dissolve</u> — e.g. hydrocarbons, oils and fats. This makes ethanol, methanol and propanol <u>very useful solvents</u> in industry.

2) <u>Ethanol</u> is the solvent for <u>perfumes</u> and <u>aftershave</u> lotions.
It can mix with both the <u>oils</u> (which give the smell) <u>and</u> the <u>water</u> (that makes up the bulk).

3) 'Methylated spirit' (or 'meths') is <u>ethanol</u> with chemicals (e.g. methanol) added to it.
It's used to <u>clean</u> paint brushes and as a <u>fuel</u> (among other things).
It's <u>poisonous</u> to drink, so a <u>purply-blue dye</u> is also added (to stop people drinking it by mistake).

Alcohols *are Used as Fuels*

1) Ethanol is used as a fuel in <u>spirit burners</u> — it burns fairly cleanly and it's non-smelly.

2) Ethanol can also be mixed in with petrol and used as <u>fuel for cars</u>. Since pure ethanol is <u>clean burning</u>, the more ethanol in a petrol/ethanol mix, the less <u>pollution</u> is produced.

3) Some countries that have little or no oil deposits but plenty of land and sunshine (e.g. Brazil) grow loads of <u>sugar cane</u>, which they <u>ferment</u> to form ethanol.

4) A big advantage of this is that sugar cane is a <u>renewable resource</u> (unlike petrol, which will run out).

Quick tip — don't fill your car with single malt whisky...

It's the <u>-OH functional group</u> that makes an alcohol an alcohol. This page is full to the brim with information about their <u>formulas</u>, <u>structures</u>, <u>properties</u> and the fantastic <u>reactions</u> you can do with them. It's a gift.

Carboxylic Acids

So what if carboxylic is a funny name — these are easy.

Carboxylic Acids Have the Functional Group -COOH

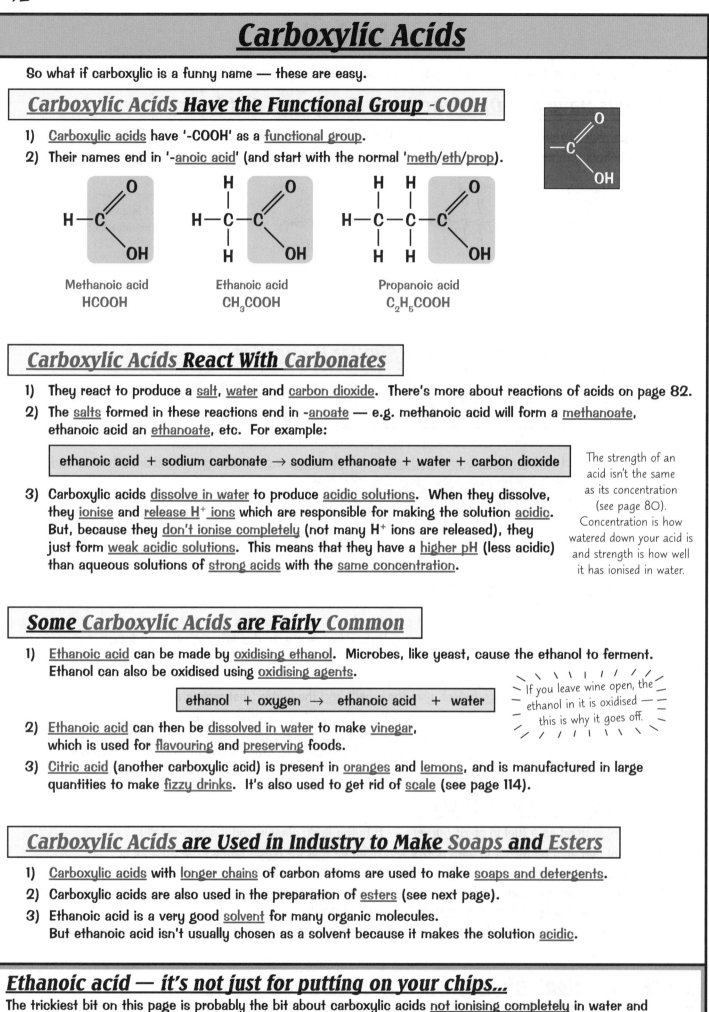

1) Carboxylic acids have '-COOH' as a functional group.

2) Their names end in '-anoic acid' (and start with the normal 'meth/eth/prop).

Methanoic acid
HCOOH

Ethanoic acid
CH_3COOH

Propanoic acid
C_2H_5COOH

Carboxylic Acids React With Carbonates

1) They react to produce a salt, water and carbon dioxide. There's more about reactions of acids on page 82.

2) The salts formed in these reactions end in -anoate — e.g. methanoic acid will form a methanoate, ethanoic acid an ethanoate, etc. For example:

ethanoic acid + sodium carbonate → sodium ethanoate + water + carbon dioxide

3) Carboxylic acids dissolve in water to produce acidic solutions. When they dissolve, they ionise and release H^+ ions which are responsible for making the solution acidic. But, because they don't ionise completely (not many H^+ ions are released), they just form weak acidic solutions. This means that they have a higher pH (less acidic) than aqueous solutions of strong acids with the same concentration.

The strength of an acid isn't the same as its concentration (see page 80). Concentration is how watered down your acid is and strength is how well it has ionised in water.

Some Carboxylic Acids are Fairly Common

1) Ethanoic acid can be made by oxidising ethanol. Microbes, like yeast, cause the ethanol to ferment. Ethanol can also be oxidised using oxidising agents.

ethanol + oxygen → ethanoic acid + water

If you leave wine open, the ethanol in it is oxidised — this is why it goes off.

2) Ethanoic acid can then be dissolved in water to make vinegar, which is used for flavouring and preserving foods.

3) Citric acid (another carboxylic acid) is present in oranges and lemons, and is manufactured in large quantities to make fizzy drinks. It's also used to get rid of scale (see page 114).

Carboxylic Acids are Used in Industry to Make Soaps and Esters

1) Carboxylic acids with longer chains of carbon atoms are used to make soaps and detergents.

2) Carboxylic acids are also used in the preparation of esters (see next page).

3) Ethanoic acid is a very good solvent for many organic molecules. But ethanoic acid isn't usually chosen as a solvent because it makes the solution acidic.

Ethanoic acid — it's not just for putting on your chips...

The trickiest bit on this page is probably the bit about carboxylic acids not ionising completely in water and being weak acids. But when it comes to carbonates, they act just like any acid (see page 82).

Esters

Mix an alcohol from p. 41 and a carboxylic acid from p. 42, and what have you got... an <u>ester</u>, that's what.

Esters *Have the Functional Group* -COO-

1) <u>Esters</u> are formed from an <u>alcohol</u> and a <u>carboxylic acid</u>.

2) An <u>acid catalyst</u> is usually used (e.g. concentrated <u>sulfuric acid</u>).

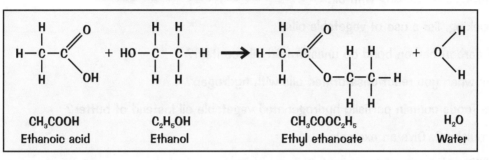

alcohol + carboxylic acid → ester + water

CH_3COOH C_2H_5OH $CH_3COOC_2H_5$ H_2O
Ethanoic acid   Ethanol   Ethyl ethanoate   Water

Their names end in '-<u>oate</u>'.
The <u>alcohol</u> forms the <u>first</u> part of the ester's
name, and the <u>acid</u> forms the <u>second</u> part.

ethanol + ethanoic acid → ethyl ethanoate + water
methanol + propanoic acid → methyl propanoate + water

Esters *Smell Nice but Don't Mix Well with Water*

1) Many esters have <u>pleasant smells</u> — often quite <u>sweet and fruity</u>. They're also <u>volatile</u> (see page 15). This makes them ideal for perfumes (the evaporated molecules can be detected by smell receptors in your nose).

2) However, many esters are <u>flammable</u> (or even <u>highly</u> flammable). So their volatility also makes them potentially <u>dangerous</u>.

3) Esters <u>don't mix very well with water</u>. (They're not nearly as soluble as alcohols or carboxylic acids.)

4) But esters do mix well with <u>alcohols</u> and other <u>organic solvents</u>.

Esters *are Often Used in* Flavourings *and* Perfumes

1) Because many esters smell nice, they're used in <u>perfumes</u>.

2) Esters are also used to make <u>flavourings</u> and <u>aromas</u> — e.g. there are esters that smell or taste of rum, apple, orange, banana, grape, pineapple, etc.

3) Some esters are used in <u>ointments</u> (they give Deep Heat® its smell).

4) Other esters are used as <u>solvents</u> for paint, ink, glue and in nail varnish remover.

There are things you need to think about when using esters:

1) Inhaling the <u>fumes</u> from some esters <u>irritates mucous membranes</u> in the nose and mouth.

2) <u>Ester fumes</u> are <u>heavier than air</u> and very <u>flammable</u>. Flammable vapour + naked flame = <u>flash fire</u>.

3) Some esters are <u>toxic</u>, especially in large doses. Some people worry about <u>health problems</u> associated with <u>synthetic food additives</u> such as esters.

4) BUT... esters <u>aren't as volatile</u> or as <u>toxic</u> as some other <u>organic solvents</u> — they don't release nearly as many toxic fumes as some of them. In fact esters have <u>replaced solvents</u> such as toluene in many paints and varnishes.

What's a chemist's favourite chocolate — ester eggs...

Turns out your favourite <u>perfume</u> is probably full of esters. Ester fumes are flammable which is why it probably says, 'Don't spray near a naked flame', or 'Keep away from sources of ignition', on the back. There you go.

Revision Summary for Section Four

The only way that you can tell if you know what you need to know is to test yourself. Try these questions and if there's something you don't know, it means you should go back and learn it. Even if it is all that weird stuff about emulsifiers. Just think of it as increasing your knowledge of mayonnaise. After all, who wouldn't want that.

1) Why do some oils need to be distilled after they have been extracted?

2) List two advantages of cooking with oil.

3) Apart from cooking, list a use of vegetable oils.

4) What kind of carbon-carbon bond do unsaturated oils contain?

5) What happens when you react unsaturated oils with hydrogen?

6) Why do some foods contain partially hydrogenated vegetable oil instead of butter?

7) What is an emulsion? Give an example.

8) How do emulsifiers keep emulsions stable?

9) Suggest one problem of adding emulsifiers to food.

10) Draw the structure of the first three alcohols.

11) When alcohols dissolve in water, is the solution acidic, neutral or alkaline?

12) What gas is formed when alcohols react with sodium?

13) Give two uses of alcohols.

14) What is the functional group in carboxylic acids?

15) Give two uses of carboxylic acids.

16) What two kinds of substance react together to form an ester?
    What catalyst is used in the formation of esters?

17) Write down two uses of esters.

The Earth's Structure

This page is all about the <u>structure</u> of <u>the Earth</u> — what the planet's like inside, and how scientists study it...

The Earth has a Crust, a Mantle and a Core

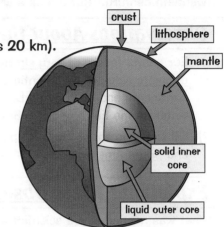

1) The <u>crust</u> is Earth's thin outer layer of solid rock (its average depth is 20 km).

2) The <u>lithosphere</u> includes the crust and upper part of the <u>mantle</u>, and is made up of a <u>jigsaw</u> of 'tectonic plates'. The <u>lithosphere</u> is <u>relatively cold and rigid</u>, and is over 100 km thick in places.

3) The <u>mantle</u> is the <u>solid</u> section between the crust and the core. Near the crust it's <u>very rigid</u>. As you go deeper into the mantle the <u>temperature increases</u> — here it becomes <u>less rigid</u> and can <u>flow very slowly</u> (it behaves like it's semi-liquid).

4) At the centre of the Earth is the <u>core</u>, which we think is made of <u>iron and nickel</u>.

5) The <u>core</u> is just over <u>half</u> the Earth's radius. The <u>inner core</u> is <u>solid</u>, while the <u>outer core</u> is <u>liquid</u>.

6) <u>Radioactive decay</u> creates a lot of the <u>heat</u> inside the Earth. This heat creates <u>convection currents</u> in the mantle, which causes the <u>plates</u> of the lithosphere to <u>move</u>.

The Earth's Surface is Made Up of Tectonic Plates

1) The crust and the upper part of the mantle are cracked into a number of large pieces called <u>tectonic plates</u>. These plates are a bit like <u>big rafts</u> that 'float' on the mantle.

2) The plates don't stay in one place though. That's because the <u>convection currents</u> in the mantle cause the plates to <u>drift</u>.

3) The map shows the <u>edges</u> of the plates as they are now, and the <u>directions</u> they're moving in (red arrows).

4) Most of the plates are moving at speeds of <u>a few cm per year</u> relative to each other.

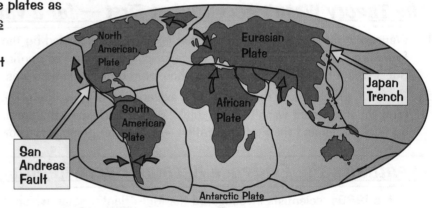

5) Occasionally, the plates move very <u>suddenly</u>, causing an <u>earthquake</u>.

6) <u>Volcanoes</u> and <u>earthquakes</u> often occur at the boundaries between two tectonic plates.

Scientists Can't Predict Earthquakes and Volcanic Eruptions

1) Tectonic plates can stay more or less put for a while and then <u>suddenly</u> lurch forwards. It's <u>impossible to predict</u> exactly when they'll move.

2) Scientists are trying to find out if there are any <u>clues</u> that an earthquake might happen soon — things like strain in underground rocks. Even with these clues they'll only be able to say an earthquake's <u>likely</u> to happen, not <u>exactly when</u> it'll happen.

3) There are some <u>clues</u> that say a volcanic eruption might happen soon. There's more about this on page 47.

2.5 cm a year — that's as fast as your fingernails grow...

So everyone standing on the surface of our little blue-green planet is actually <u>floating</u> round very slowly on a sea of <u>semi-liquid rock</u>. Make sure you understand the stuff about <u>tectonic plates</u> — there's more coming up...

Plate Tectonics

The idea that the Earth's surface is made up of moving plates of rock has been around since the early twentieth century. But it took a while to catch on.

Observations About the Earth Hadn't Been Explained

1) For years, fossils of very similar plants and animals had been found on opposite sides of the Atlantic Ocean. Most people thought this was because the continents had been linked by 'land bridges', which had sunk or been covered by water as the Earth cooled. But not everyone was convinced, even back then.

2) Other things about the Earth puzzled people too — like why the coastlines of Africa and South America fit together and why there are fossils of sea creatures in the Alps.

Explaining These Observations Needed a Leap of Imagination

What was needed was a scientist with a bit of insight... a smidgeon of creativity... a touch of genius...

1) In 1914 Alfred Wegener hypothesised that Africa and South America had previously been one continent which had then split. He started to look for more evidence to back up his hypothesis. He found it...

2) E.g. there were matching layers in the rocks on different continents, and similar earthworms living in both South America and South Africa.

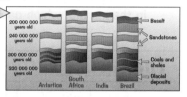

3) Wegener's theory of 'continental drift' supposed that about 300 million years ago there had been just one 'supercontinent' — which he called Pangaea. According to Wegener, Pangaea broke into smaller chunks, and these chunks (our modern-day continents) are still slowly 'drifting' apart. This idea is the basis behind the modern theory of plate tectonics.

The Theory Wasn't Accepted at First — for a Variety of Reasons

1) Wegener's theory explained things that couldn't be explained by the 'land bridge' theory (e.g. the formation of mountains — which Wegener said happened as continents smashed into each other). But it was a big change, and the reaction from other scientists was hostile.

2) The main problem was that Wegener's explanation of how the 'drifting' happened wasn't convincing (and the movement wasn't detectable). Wegener claimed the continents' movement could be caused by tidal forces and the Earth's rotation — but other geologists showed that this was impossible.

Eventually, the Evidence Became Overwhelming

1) In the 1960s, scientists investigated the Mid-Atlantic ridge, which runs the whole length of the Atlantic.

2) They found evidence that magma (molten rock) rises up through the sea floor, solidifies and forms underwater mountains that are roughly symmetrical either side of the ridge. The evidence suggested that the sea floor was spreading — at about 10 cm per year.

3) Even better evidence that the continents are moving apart came from the magnetic orientation of the rocks. As the liquid magma erupts out of the gap, iron particles in the rocks tend to align themselves with the Earth's magnetic field — and as it cools they set in position. Now then... every half million years or so the Earth's magnetic field swaps direction — and the rock on either side of the ridge has bands of alternate magnetic polarity, symmetrical about the ridge.

4) This was convincing evidence that new sea floor was being created... and continents were moving apart.

5) All the evidence collected by other scientists supported Wegener's theory — so it was gradually accepted.

I told you so — but no one ever believes me...

Wegener wasn't right about everything, but his main idea was correct. The scientific community was a bit slow to accept it, but once there was more evidence to support it, they got on board. That's science for you...

Volcanic Eruptions

The theory of <u>plate tectonics</u> not only explains why the continents move, it also makes sense of natural hazards such as <u>volcanoes</u> and <u>earthquakes</u>.

Volcanoes are Formed by Molten Rock

1) Volcanoes occur when <u>molten rock</u> (<u>magma</u>) from the <u>mantle</u> emerges through the Earth's crust.

2) Magma rises up (through the crust) and 'boils over' where it can — sometimes quite violently if the pressure is released suddenly. (When the molten rock is <u>below</u> the surface of the Earth it's called <u>magma</u> — but when it <u>erupts</u> from a volcano it's called <u>lava</u>.)

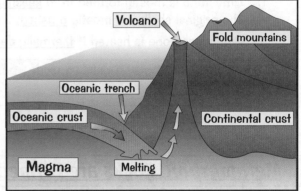

Oceanic and Continental Crust Colliding Causes Volcanoes

1) The crust at the <u>ocean floor</u> is <u>denser</u> than the crust below the <u>continents</u>.

2) When two tectonic plates collide, a dense <u>oceanic plate</u> will be <u>forced underneath</u> a less dense <u>continental plate</u>. This is called <u>subduction</u>.

3) Oceanic crust also tends to be <u>cooler</u> at the <u>edges</u> of a tectonic plate — so the edges <u>sink</u> easily, <u>pulling</u> the oceanic plate <u>down</u>.

4) As the oceanic crust is forced down it <u>melts</u> and <u>starts to rise</u>. If this <u>molten rock</u> finds its way to the <u>surface</u>, <u>volcanoes</u> form.

Volcano

Fold mountains

Oceanic trench

Oceanic crust

Continental crust

Magma

Melting

Volcanic Activity Forms Igneous Rock

1) <u>Igneous rock</u> is made when any sort of <u>molten rock cools down</u> and <u>solidifies</u>. Lots of rocks on the surface of the Earth were formed this way.

2) The <u>type</u> of igneous rock (and the behaviour of the volcano) depends on how <u>quickly</u> the magma <u>cools</u> and the <u>composition of the magma</u>.

3) Some volcanoes produce magma that forms <u>iron-rich basalt</u>. The lava from the eruption is runny, and the eruption is <u>fairly safe</u>. (As safe as you can be with molten rock at 1200 °C, I suppose.)

4) But if the magma is <u>silica-rich rhyolite</u>, the eruption is <u>explosive</u>. It produces <u>thick lava</u> which can be violently blown out of the top of the volcano. Crikey.

Geologists Try to Predict Volcanic Eruptions

1) Geologists study volcanoes to try to find out if there are <u>signs</u> that a volcanic eruption might happen soon — things like <u>magma movement</u> below the ground near to a volcano. This causes <u>mini-earthquakes</u> near the volcano.

2) Being able to spot these kinds of clues means that scientists can <u>predict eruptions</u> with much <u>greater accuracy</u> than they could in the past.

3) It's tricky though — volcanoes are very unpredictable. For example, sometimes molten rock cools down instead of erupting, so mini-earthquakes can be a <u>false alarm</u>.

4) Most likely, scientists will only be able to say that an eruption's <u>more likely than normal</u> — not that it's <u>certain</u>. But even just knowing that can <u>save lives</u>.

__Make the Earth move for you — stand next to a volcano...__

Volcanoes can erupt with huge force, so it might seem odd to choose to <u>live</u> near one. But there are <u>benefits</u> — volcanic ash creates very <u>fertile soil</u> that's great for farming. It would be much <u>safer</u> if eruptions could be <u>predicted</u> accurately — geologists aren't there yet, but their predictions are getting better all the time.

The Three Different Types of Rock

Scientists classify rocks according to how they're formed. The three different types are: <u>sedimentary</u>, <u>metamorphic</u> and <u>igneous</u>. Sedimentary rocks are generally pretty soft, while igneous rocks are well hard.

There are *Three Steps* in the Formation of *Sedimentary Rock*

1) <u>Sedimentary rocks</u> are formed from <u>layers of sediment</u> laid down in <u>lakes</u> or <u>seas</u>.

2) Over <u>millions of years</u> the layers get <u>buried</u> under more layers and the <u>weight</u> pressing down <u>squeezes out</u> the water.

3) Fluids flowing through the pores deposit natural mineral <u>cement</u>.

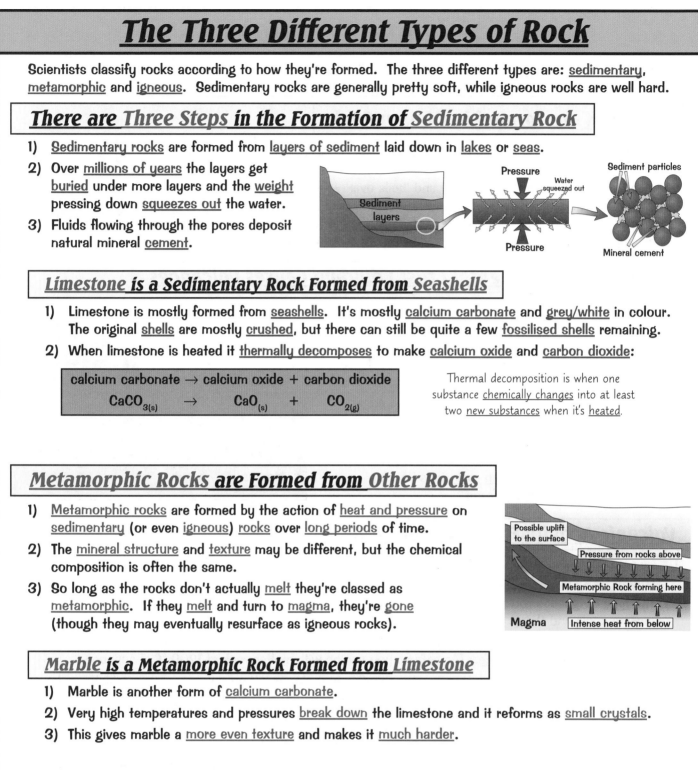

Limestone is a Sedimentary Rock Formed from *Seashells*

1) Limestone is mostly formed from <u>seashells</u>. It's mostly <u>calcium carbonate</u> and <u>grey/white</u> in colour. The original <u>shells</u> are mostly <u>crushed</u>, but there can still be quite a few <u>fossilised shells</u> remaining.

2) When limestone is heated it <u>thermally decomposes</u> to make <u>calcium oxide</u> and <u>carbon dioxide</u>:

calcium carbonate → calcium oxide + carbon dioxide
$$CaCO_{3(s)} \rightarrow CaO_{(s)} + CO_{2(g)}$$

Thermal decomposition is when one substance <u>chemically changes</u> into at least two <u>new substances</u> when it's <u>heated</u>.

Metamorphic Rocks are Formed from Other Rocks

1) <u>Metamorphic rocks</u> are formed by the action of <u>heat and pressure</u> on <u>sedimentary</u> (or even <u>igneous</u>) rocks over <u>long periods</u> of time.

2) The <u>mineral structure</u> and <u>texture</u> may be different, but the chemical composition is often the same.

3) So long as the rocks don't actually <u>melt</u> they're classed as <u>metamorphic</u>. If they <u>melt</u> and turn to <u>magma</u>, they're <u>gone</u> (though they may eventually resurface as igneous rocks).

Marble is a Metamorphic Rock Formed from *Limestone*

1) Marble is another form of <u>calcium carbonate</u>.

2) Very high temperatures and pressures <u>break down</u> the limestone and it reforms as <u>small crystals</u>.

3) This gives marble a <u>more even texture</u> and makes it <u>much harder</u>.

Igneous Rocks are Formed from *Fresh Magma*

1) <u>Igneous rocks</u> are formed when <u>magma</u> cools (see previous page).

2) They contain various <u>different minerals</u> in <u>randomly arranged</u> interlocking <u>crystals</u> — this makes them very <u>hard</u>.

3) Granite is a <u>very hard</u> igneous rock (even harder than marble). It's ideal for <u>steps</u> and <u>buildings</u>.

Igneous rocks are real cool — or they're magma...

There are a few scientific terms on this page to get your teeth into, but there's nothing too tricky to get your head round. Just remember that <u>limestone</u> is a <u>sedimentary rock</u>, <u>marble</u> is a <u>metamorphic rock</u> and <u>granite</u> is an <u>igneous rock</u>. And that's why granite is harder than marble, which is harder than limestone. Job's a good 'un.

The Evolution of the Atmosphere

For 200 million years or so, the atmosphere has been about how it is now: approximately 78% nitrogen, 21% oxygen, 1% argon and small amounts of other gases, mainly carbon dioxide, noble gases and water vapour. But it wasn't always like this. Here's how the past 4.5 billion years may have gone:

Phase 1 — Volcanoes Gave Out Gases

1) The Earth's surface was originally <u>molten</u> for many millions of years. It was so hot that any atmosphere just '<u>boiled away</u>' into space.

2) Eventually things cooled down a bit and a <u>thin crust</u> formed, but <u>volcanoes</u> kept erupting.

3) The volcanoes gave out lots of gas. We think this was how the oceans and atmosphere were formed.

4) The early atmosphere was probably <u>mostly CO₂</u>, with virtually <u>no oxygen</u>. There may also have been <u>water vapour</u>, and small amounts of <u>methane</u> and <u>ammonia</u>. This is quite like the atmospheres of Mars and Venus today.

5) The <u>oceans</u> formed when the water vapour <u>condensed</u>.

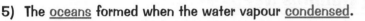

<u>Holiday report</u>: Not a nice place to be. Take strong walking boots and a good coat.

Phase 2 — Green Plants Evolved and Produced Oxygen

<u>Holiday report</u>: A bit slimy underfoot. Take wellies and a lot of suncream.

1) <u>Green plants</u> and <u>algae</u> evolved over most of the Earth. They were quite happy in the <u>CO₂ atmosphere</u>.

2) A lot of the early CO₂ <u>dissolved</u> into the oceans. The <u>green plants</u> and <u>algae</u> also absorbed some of the <u>CO₂</u> and <u>produced O₂</u> by <u>photosynthesis</u>.

3) Plants and algae died and were buried under layers of sediment, along with the skeletons and shells of marine organisms that had slowly evolved. The <u>carbon</u> and <u>hydrocarbons</u> inside them became 'locked up' in <u>sedimentary rocks</u> as <u>insoluble carbonates</u> (e.g. limestone) and <u>fossil fuels</u>.

4) When we <u>burn</u> fossil fuels today, this 'locked-up' carbon is released and the concentration of CO₂ in the atmosphere rises.

Phase 3 — Ozone Layer Allows Evolution of Complex Animals

1) The build-up of <u>oxygen</u> in the atmosphere <u>killed off</u> some early organisms that couldn't tolerate it, but allowed other, more complex organisms to evolve and flourish.

2) The oxygen also created the <u>ozone layer</u> (O₃) which <u>blocked</u> harmful rays from the Sun and <u>enabled</u> even <u>more complex</u> organisms to evolve — us, eventually.

3) There is virtually <u>no CO₂</u> left now.

<u>Holiday report</u>: A nice place to be. Visit before the crowds ruin it.

The atmosphere's evolving — shut the window will you...

We've learned a lot about the past atmosphere from <u>Antarctic ice cores</u>. Each year, a layer of ice forms and <u>bubbles of air</u> get trapped inside it, then it's buried by the next layer. So the deeper the ice, the older the air — and if you examine the bubbles in different layers, you can see how the air has changed.

Life, Resources and Atmospheric Change

Life on Earth began <u>billions of years</u> ago, but there's no way of knowing for definite how it all started.

Primordial Soup is Just One Theory of How Life was Formed

1) The primordial soup theory states that billions of years ago, the Earth's <u>atmosphere</u> was rich in <u>nitrogen</u>, <u>hydrogen</u>, <u>ammonia</u> and <u>methane</u>.

2) <u>Lightning</u> struck, causing a chemical reaction between the gases, resulting in the formation of <u>amino acids</u>.

3) The amino acids collected in a '<u>primordial soup</u>' — a body of water out of which life gradually crawled.

4) The amino acids gradually combined to produce <u>organic matter</u> which eventually evolved into simple <u>living organisms</u>.

5) In the 1950s, <u>Miller and Urey</u> carried out an experiment to prove this theory. They sealed the gases in their apparatus, heated them and applied an electrical charge for a week.

6) They found that <u>amino acids were made</u>, but not as many as there are on Earth. This suggests the theory could be along the <u>right lines</u>, but isn't quite right.

The Earth Has All the Resources Humans Need

The Earth's crust, oceans and atmosphere are the <u>ultimate source</u> of minerals and resources — we can get everything we need from them. For example, we can <u>fractionally distil air</u> to get a variety of products (e.g. nitrogen and oxygen) for use in <u>industry</u>:

1) Air is <u>filtered</u> to remove dust.

2) It's then <u>cooled</u> to around <u>-200 °C</u> and becomes a liquid.

3) During cooling <u>water vapour</u> condenses and is removed.

4) <u>Carbon dioxide</u> freezes and is removed.

5) The liquefied air then enters the fractionating column and is <u>heated</u> slowly.

6) The remaining gases are separated by <u>fractional distillation</u>. Oxygen and argon come out together so <u>another</u> column is used to separate them.

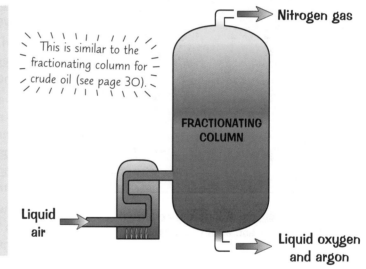

This is similar to the fractionating column for crude oil (see page 30).

FRACTIONATING COLUMN

Nitrogen gas

Liquid air

Liquid oxygen and argon

Increasing Carbon Dioxide Level Affects the Climate and the Oceans

<u>Burning fossil fuels</u> releases CO_2 — and as the world's become more industrialised, more fossil fuels have been burnt in power stations and in car engines. This CO_2 is thought to be altering our planet...

1) An increase in carbon dioxide is causing <u>global warming</u> — a type of <u>climate change</u> (see page 53).

2) The oceans are a <u>natural store</u> of CO_2 — they absorb it from the atmosphere. However the extra CO_2 we're releasing is making them too <u>acidic</u>. This is bad news for <u>coral</u> and <u>shellfish</u>, and also means that in the future they won't be able to absorb any more carbon dioxide.

Waiter, waiter, there's a primate in my soup...

No-one was around billions of years ago, so our theories about how life formed are just that — theories. We're also still guessing about the exact effects of global warming on things like the oceans.

Air Pollution — Carbon and Sulfur

We burn fuels all the time to release the energy stored inside them — e.g. 90% of crude oil is used as fuel.

Burning Fossil Fuels Releases Gases and Particles

1) Power stations burn huge amounts of fossil fuels to make electricity. Cars are also a major culprit in burning fossil fuels.

2) Most fuels, such as crude oil and coal, contain carbon and hydrogen. During combustion, the carbon and hydrogen are oxidised so that carbon dioxide and water vapour are released into the atmosphere. Energy (heat) is also produced.
E.g.:

   hydrocarbon + oxygen → carbon dioxide + water vapour

3) When there's plenty of oxygen, all the fuel burns — this is called complete combustion.

4) If there's not enough oxygen, some of the fuel doesn't burn — this is called partial combustion. Under these conditions, solid particles (called particulates) of soot (carbon) and unburnt fuel are released. Carbon monoxide (a poisonous gas) is also released.

Partial combustion is also known as incomplete combustion.

Sulfur Dioxide Causes Acid Rain

1) Sulfur dioxide is one of the gases that causes acid rain.

2) When the sulfur dioxide mixes with clouds it forms dilute sulfuric acid. This then falls as acid rain.

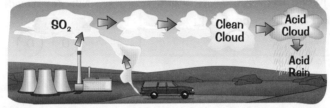

3) In the same way, oxides of nitrogen cause acid rain by forming dilute nitric acid in clouds.

4) Acid rain causes lakes to become acidic and many plants and animals die as a result.

5) Acid rain kills trees and damages limestone buildings and ruins stone statues. It's shocking.

6) Links between acid rain and human health problems have been suggested.

7) The benefits of electricity and travel have to be balanced against the environmental impacts. Governments have recognised the importance of this and international agreements have been put in place to reduce emissions of air pollutants such as sulfur dioxide.

You can Reduce Acid Rain by Reducing Sulfur Emissions

1) Most of the sulfur can be removed from fuels before they're burnt, but it costs more to do it.

2) Also, removing sulfur from fuels takes more energy. This usually comes from burning more fuel, which releases more of the greenhouse gas carbon dioxide.

3) However, petrol and diesel are starting to be replaced by low-sulfur versions.

4) Power stations now have Acid Gas Scrubbers to take the harmful gases out before they release their fumes into the atmosphere.

5) The other way of reducing acid rain is simply to reduce our usage of fossil fuels.

Eee, problems, problems — there's always summat goin' wrong...

Pollutants like sulfur dioxide can be carried a long way in the atmosphere. So a country might suffer from acid rain that it didn't cause, which doesn't seem very fair. It's not just up to big industries though — there's lots of things you can do to reduce the amount of fossil fuels burnt. Putting an extra jumper on instead of turning up the heating helps. As does walking places instead of cadging a lift.

Air Pollution — Nitrogen

Now for a new type of pollution. One that, surprisingly, is made from the nitrogen in the air itself.

Nitrogen Pollution Involves Nitrogen from the Air

Nitrogen pollution doesn't actually come from the fuel itself — it's formed from nitrogen in the air when the fuel is burnt.

1) Fossil fuels burn at such high temperatures that nearby atoms in the air react with each other.
2) Nitrogen in the air reacts with the oxygen in the air to produce small amounts of compounds known as nitrogen oxides — nitrogen monoxide and nitrogen dioxide.
3) This happens in car engines.
4) Nitrogen oxides are pollutants, and are usually spewed straight out into the atmosphere.

Nitrogen Oxides Are Nitrogen Monoxide and Nitrogen Dioxide

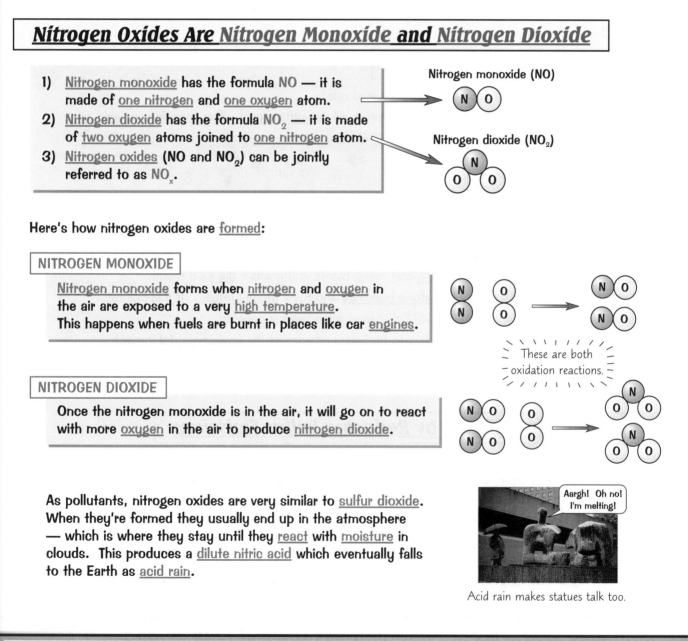

1) Nitrogen monoxide has the formula NO — it is made of one nitrogen and one oxygen atom.
2) Nitrogen dioxide has the formula NO_2 — it is made of two oxygen atoms joined to one nitrogen atom.
3) Nitrogen oxides (NO and NO_2) can be jointly referred to as NO_x.

Nitrogen monoxide (NO)

Nitrogen dioxide (NO_2)

Here's how nitrogen oxides are formed:

NITROGEN MONOXIDE

Nitrogen monoxide forms when nitrogen and oxygen in the air are exposed to a very high temperature. This happens when fuels are burnt in places like car engines.

These are both oxidation reactions.

NITROGEN DIOXIDE

Once the nitrogen monoxide is in the air, it will go on to react with more oxygen in the air to produce nitrogen dioxide.

As pollutants, nitrogen oxides are very similar to sulfur dioxide. When they're formed they usually end up in the atmosphere — which is where they stay until they react with moisture in clouds. This produces a dilute nitric acid which eventually falls to the Earth as acid rain.

Aargh! Oh no! I'm melting!

Acid rain makes statues talk too.

What we need is a pollution solution...

Oh good — another type of pollution to worry about. Well, in my opinion, the only answer is to go back to the old days of farming a bit of land in a self-sufficient manner. No more electricity, no more industry, bliss. Although there'd be no more TVs, DVD players, X-ray machines, computers. Hmm, in that case...

Environmental Problems

More doom and gloom on this page I'm afraid...

Increasing Carbon Dioxide Causes Climate Change

1) The level of carbon dioxide in the atmosphere is increasing — because of the large amounts of fossil fuels humans burn.

2) There's a scientific consensus that this extra carbon dioxide has caused the average temperature of the Earth to increase — global warming.

3) Global warming is a type of climate change and causes other types of climate change, e.g. changing rainfall patterns. It could also cause severe flooding due to the polar ice caps melting.

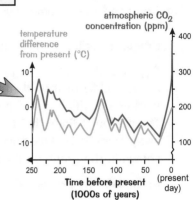

Particles Cause Global Dimming

1) In the last few years, some scientists have been measuring how much sunlight is reaching the surface of the Earth and comparing it to records from the last 50 years.

2) They have been amazed to find that in some areas nearly 25% less sunlight has been reaching the surface compared to 50 years ago. They have called this global dimming.

3) They think that it is caused by particles of soot and ash that are produced when fossil fuels are burnt. These particles reflect sunlight back into space, or they can help to produce more clouds that reflect the sunlight back into space.

4) There are many scientists who don't believe the change is real and blame it on inaccurate recording equipment.

Alternative Fuels are Being Developed

Some alternative fuels have already been developed, and there are others in the pipeline (so to speak). Many of them are renewable fuels so, unlike fossil fuels, they won't run out. However, none of them are perfect — they all have pros and cons. For example:

ETHANOL can be produced from plant material so is known as a biofuel. It's made by fermentation of plants and is used to power cars in some places. It's often mixed with petrol to make a better fuel.

PROS: The CO_2 released when it's burnt was taken in by the plant as it grew, so it's 'carbon neutral'. The only other product is water.

CONS: Engines need to be converted before they'll work with ethanol fuels. And ethanol fuel isn't widely available. There are worries that as demand for it increases farmers will switch from growing food crops to growing crops to make ethanol — this will increase food prices.

BIODIESEL is another type of biofuel. It can be produced from vegetable oils such as rapeseed oil and soybean oil. Biodiesel can be mixed with ordinary diesel fuel and used to run a diesel engine.

PROS: Biodiesel is 'carbon neutral'. Engines don't need to be converted. It produces much less sulfur dioxide and 'particulates' than ordinary diesel or petrol.

CONS: We can't make enough to completely replace diesel. It's expensive to make. It could increase food prices like using more ethanol could (see above).

HYDROGEN GAS can also be used to power vehicles. You get the hydrogen from the electrolysis of water — there's plenty of water about but it takes electrical energy to split it up. This energy can come from a renewable source, e.g. solar.

PROS: Hydrogen combines with oxygen in the air to form just water — so it's very clean.

CONS: You need a special, expensive engine and hydrogen isn't widely available. You still need to use energy from another source to make it. Also, hydrogen's hard to store.

Global dimming — romantic lighting all day...

Alternative fuels are the shining light at the end of a long tunnel of problems caused by burning fuels (and I mean long). But nothing's perfect — except my quiff... and maybe my golf swing... and maybe Cilla Black.

Revision Summary for Section Five

These certainly aren't the easiest questions you're going to come across. That's because they test what you know without giving you any clues. At first you might think they're impossibly difficult. Eventually you'll realise that they simply test whether you've learnt the stuff or not. If you're struggling to answer these then you need to do some serious learning.

1) Briefly describe the inner structure of the Earth.

2) A geologist places a very heavy marker on the seabed in the middle of the Atlantic ocean. She records the marker's position over a period of four years. The geologist finds that the marker moves in a straight-line away from its original position. Her measurements are shown in the graph on the right.

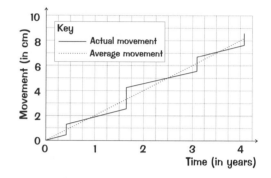

a) Explain the process that has caused the marker to move.

b)*What is the marker's average movement each year?

c)*On average, how many years will it take for the marker to move 7 cm?

3) Why can't scientists accurately predict volcanoes and earthquakes?

4) Describe the evidence that backs up Wegener's theory of continental drift.

5) What is meant by 'subduction'?

6) Sketch a labelled diagram showing how a volcano forms.

7) How is an eruption of silica-rich rhyolitic lava different from an eruption of iron-rich basaltic lava?

8) Draw a diagram to show how metamorphic rocks form.

9) Give an example of a metamorphic rock and say what the material it formed from is.

10) Which material is hardest, granite, limestone or marble?

11) Name the three main gases that make up the Earth's atmosphere today.

12) Explain why today's atmosphere is different from the Earth's early atmosphere.

13) What is meant by 'primordial soup'?

14) Why do we fractionally distil air?

15) The burning of fossils fuels is causing a rise in the level of carbon dioxide in the atmosphere. How is this affecting the oceans and the climate?

16) Describe briefly how the pollutant sulfur dioxide is produced.

17) What effects does acid rain have on the environment?

18) List three ways of reducing acid rain.

19) Describe how nitrogen dioxide is produced when fossil fuels are burnt.

20) Give the formulas of nitrogen monoxide and nitrogen dioxide.

21) What effect do nitrogen oxides have on the environment?

22) Has the theory of global dimming been proven?

23) List three alternative ways of powering cars. What are the pros and cons of each?

* Answers on page 140.

History of the Periodic Table

We haven't always known as much about Chemistry as we do now. No sirree. Early chemists looked to try and understand patterns in the elements' properties to get a bit of understanding.

Döbereiner Tried to Organise Elements into Triads

Back in the 1800s the only thing they could measure was relative atomic mass, and so the known elements were arranged in order of atomic mass.

In 1828 a guy called Döbereiner started to put this list of elements into groups based on their chemical properties. He put the elements into groups of three, which he called triads. E.g. Cl, Br and I were one triad, and Li, Na and K were another.

The middle element of each triad had a relative atomic mass that was the average of the other two.

Element	Relative atomic mass
Lithium	7
Sodium	23
Potassium	39

$(7 + 39) \div 2 = 23$

Newlands' Law of Octaves Was the First Good Effort

A chap called Newlands noticed that when you arranged the elements in order of relative atomic mass, every eighth element had similar properties, and so he listed some of the known elements in rows of seven:

H	Li	Be	B	C	N	O
F	Na	Mg	Al	Si	P	S
Cl	K	Ca	Cr	Ti	Mn	Fe

These sets of eight were called Newlands' Octaves. Unfortunately the pattern broke down on the third row, with transition metals like titanium (Ti) and iron (Fe) messing it up.

It was because he left no gaps that his work was ignored. But he was getting pretty close.

Newlands presented his ideas to the Chemical Society in 1865. But his work was criticised because:

1) His groups contained elements that didn't have similar properties, e.g. carbon and titanium.

2) He mixed up metals and non-metals e.g. oxygen and iron.

3) He didn't leave any gaps for elements that hadn't been discovered yet.

Dmitri Mendeleev Left Gaps and Predicted New Elements

1) In 1869, Dmitri Mendeleev in Russia, armed with about 50 known elements, arranged them into his Table of Elements — with various gaps as shown.

2) Mendeleev put the elements in order of atomic mass (like Newlands did). But Mendeleev found he had to leave gaps in order to keep elements with similar properties in the same vertical groups — and he was prepared to leave some very big gaps in the first two rows before the transition metals come in on the third row.

Mendeleev's Table of the Elements

```
H
Li  Be                                          B   C   N   O   F
Na  Mg                                          Al  Si  P   S   Cl
K   Ca  *   Ti  V   Cr  Mn  Fe  Co  Ni  Cu  Zn  *   *   As  Se  Br
Rb  Sr  Y   Zr  Nb  Mo  *   Ru  Rh  Pd  Ag  Cd  In  Sn  Sb  Te  I
Cs  Ba  *   *   Ta  W   *   Os  Ir  Pt  Au  Hg  Tl  Pb  Bi
```

3) The gaps were the really clever bit because they predicted the properties of so far undiscovered elements. When they were found and they fitted the pattern it helped confirm Mendeleev's ideas. For example, Mendeleev made really good predictions about the chemical and physical properties of an element he called ekasilicon, which we know today as germanium.

- When the periodic table was first released, many scientists thought it was just a bit of fun. At that time, there wasn't all that much evidence to suggest that the elements really did fit together in that way.

- After Mendeleev released his work, newly discovered elements fitted into the gaps he left. This was convincing evidence in favour of the periodic table.

Julie Andrews' octaves — do-re-mi-fa-so-la-ti-do...

This is a good example of how science often works. A scientist has a basically good (though incomplete) idea. Other scientists disagree with it. Eventually, it gets modified a bit and voilà — into the textbooks it goes.

The Modern Periodic Table

Chemists were getting pretty close to producing something useful.
The big breakthrough came when the <u>structure</u> of the <u>atom</u> was understood a bit better.

Not All Scientists Thought the Periodic Table was Important

1) When the periodic table was first released, many scientists thought it was just a bit of <u>fun</u>. At that time, there wasn't all that much <u>evidence</u> to suggest that the elements really did fit together in that way — ideas don't get the scientific stamp of approval without evidence.

2) After Mendeleev released his work, <u>newly discovered elements</u> fitted into the <u>gaps</u> he left. This was convincing evidence in favour of the periodic table.

3) Once there was more evidence, many more scientists realised that the periodic table could be a <u>useful tool</u> for <u>predicting</u> properties of elements. It <u>really worked</u>.

4) In the late 19th century, scientists discovered protons, neutrons and electrons. The periodic table <u>matches up</u> very well to what's been discovered about the <u>structure</u> of the atom. Scientists now accept that it's a very important and useful <u>summary of the structure of atoms</u>.

The Modern Periodic Table is Based on Electronic Structure

When <u>electrons</u>, <u>protons</u> and <u>neutrons</u> were discovered, the periodic table was arranged in order of atomic number.
All elements were put into <u>groups</u>.

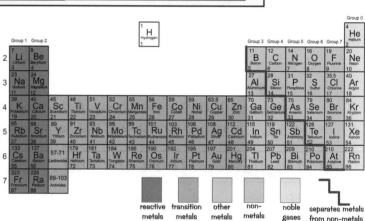

1) The elements in the periodic table can be seen as being arranged by their <u>electronic structure</u>. Using the electron arrangement, you can predict the element's <u>chemical properties</u>.

2) Electrons in an atom are set out in <u>shells</u> which each correspond to an <u>energy level</u>.

3) Apart from the transition metals, elements in the same group have the <u>same number of electrons</u> in their <u>highest occupied energy level</u> (outer shell).

4) The <u>group number</u> is <u>equal</u> to the <u>number of electrons</u> in the <u>highest occupied energy level</u> — e.g. Group 6 all have 6 electrons in the highest energy level.

5) The positive charge of the nucleus attracts electrons and holds them in place. The <u>further</u> from the nucleus the electron is, the <u>less the attraction</u>.

6) The attraction of the nucleus is <u>even less</u> when there are a lot of <u>inner electrons</u>. Inner electrons "get in the way" of the nuclear charge, reducing the attraction. This effect is known as <u>shielding</u>.

7) The combination of <u>increased distance</u> and <u>increased shielding</u> means that an electron in a higher energy level is <u>more easily lost</u> because there's <u>less attraction</u> from the nucleus holding it in place. That's why <u>Group 1 metals</u> get <u>more reactive</u> as you go down the group.

8) <u>Increased distance</u> and <u>shielding</u> also means that a higher energy level is <u>less likely to gain an electron</u> — there's less attraction from the nucleus pulling electrons into the atom. That's why <u>Group 7 elements</u> get <u>less reactive</u> going down the group.

You are now approaching Group 3 — mind the gap...

So... the <u>group number</u> of an element is <u>always</u> the number of electrons that element has in its <u>highest energy level</u>. So elements in the <u>same group</u> will have the <u>same</u> number of electrons in their highest energy level. Got it.

Isotopes and Relative Atomic Mass

Some elements have more than one isotope.

Isotopes *Are the Same Except for an Extra* Neutron *or Two*

Here's the definition of an isotope:

> Isotopes are: different atomic forms of the same element, which have the SAME number of PROTONS but a DIFFERENT number of NEUTRONS.

1) The upshot is: isotopes must have the same atomic number but different mass numbers.

2) If they had different atomic numbers, they'd be different elements altogether.

3) Carbon-12 and carbon-14 are a very popular pair of isotopes.

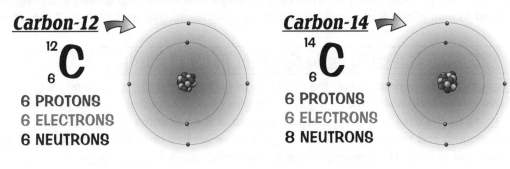

Carbon-12 $\longrightarrow$
$$^{12}_{6}\text{C}$$
6 PROTONS
6 ELECTRONS
6 NEUTRONS

Carbon-14 $\longrightarrow$
$$^{14}_{6}\text{C}$$
6 PROTONS
6 ELECTRONS
8 NEUTRONS

Relative Atomic Mass *Takes* Isotopes *into Account*

1) Relative atomic mass (A_r) uses the average mass of the isotopes of an element. It has to allow for the relative mass of each isotope and its relative abundance.

2) Relative abundance just means how much there is of each isotope compared to the total amount of the element in the world. This can be a ratio, a fraction or a percentage.

EXAMPLE: Work out the relative atomic mass of chlorine.

element	relative mass of isotope	relative abundance
chlorine	35	3
	37	1

ANSWER:

This means that there are 2 isotopes of chlorine. One has a relative mass of 35 (^{35}Cl) and the other 37 (^{37}Cl).

The relative abundances show that there are 3 atoms of ^{35}Cl to every 1 of ^{37}Cl.

- First, multiply the mass of each isotope by its relative abundance.
- Add those together.
- Divide by the sum of the relative abundances.

$$A_r = \frac{(35 \times 3) + (37 \times 1)}{3 + 1} = \underline{35.5}$$

3) Relative atomic masses don't usually come out as whole numbers or easy decimals, but they're often rounded to the nearest 0.5 in periodic tables (see page 56).

Will you ace your GCSEs — isotope so...

The thing to understand about an isotope is that it's a slight variation on the same element. Not so crazy really.

Ionic Bonding

Ionic Bonding — Transferring Electrons

In ionic bonding, atoms lose or gain electrons to form charged particles (called ions) which are then strongly attracted to one another (because of the attraction of opposite charges, + and –).

A Shell with Just One Electron is Well Keen to Get Rid...

All the atoms over at the left-hand side of the periodic table, e.g. sodium, potassium, calcium etc. have just one or two electrons in their outer shell (highest energy level). And they're pretty keen to get shot of them, because then they'll only have full shells left, which is how they like it. (They try to have the same electronic structure as a noble gas.) So given half a chance they do get rid, and that leaves the atom as an ion instead. Now ions aren't the kind of things that sit around quietly watching the world go by. They tend to leap at the first passing ion with an opposite charge and stick to it like glue.

A Nearly Full Shell is Well Keen to Get That Extra Electron...

On the other side of the periodic table, the elements in Group 6 and Group 7, such as oxygen and chlorine, have outer shells which are nearly full. They're obviously pretty keen to gain that extra one or two electrons to fill the shell up. When they do of course they become ions (you know, not the kind of things to sit around) and before you know it, pop, they've latched onto the atom (ion) that gave up the electron a moment earlier. The reaction of sodium and chlorine is a classic case:

The sodium atom gives up its outer electron and becomes an Na⁺ ion.

The chlorine atom has picked up the spare electron and becomes a Cl⁻ ion.

POP!

Ionic Compounds Have A Regular Lattice Structure

1) Ionic compounds always have giant ionic lattices.
2) The ions form a closely packed regular lattice arrangement.
3) There are very strong electrostatic forces of attraction between oppositely charged ions, in all directions.
4) A single crystal of sodium chloride (salt) is one giant ionic lattice, which is why salt crystals tend to be cuboid in shape. The Na⁺ and Cl⁻ ions are held together in a regular lattice.

● = Cl⁻
● = Na⁺

Ionic Compounds All Have Similar Properties

1) They all have high melting points and high boiling points due to the strong attraction between the ions. It takes a large amount of energy to overcome this attraction. When ionic compounds melt, the ions are free to move and they'll carry electric current.

2) They do dissolve easily in water though. The ions separate and are all free to move in the solution, so they'll carry electric current.

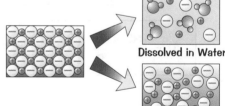

Dissolved in Water

Melted

Giant ionic lattices — all over your chips...

These guys are tough nuts to crack, but if you do crack 'em, they get all excited and start conducting electricity.

Ions and Formulas

Make sure you've really got your head around the idea of ionic bonding before you start on this page.

Groups 1 & 2 and 6 & 7 are the Most Likely to Form Ions

1) Remember, atoms that have lost or gained an electron (or electrons) are ions.

2) Ions have the electronic structure of a noble gas.

3) The elements that most readily form ions are those in Groups 1, 2, 6 and 7.

4) Group 1 and 2 elements are metals and they lose electrons to form positive ions.

5) For example, Group 1 elements (the alkali metals) form ionic compounds with non-metals where the metal ion has a 1+ charge. E.g. K^+Cl^-.

6) Group 6 and 7 elements are non-metals. They gain electrons to form negative ions.

7) For example, Group 7 elements (the halogens) form ionic compounds with the alkali metals where the halide ion has a 1− charge. E.g. Na^+Cl^-.

8) The charge on the positive ions is the same as the group number of the element.

Positive Ions		Negative Ions	
1^+ ions	2^+ ions	2^- ions	1^- ions
All Group 1 metals, including:	All Group 2 metals, including:	Carbonate CO_3^{2-}	Hydroxide OH^-
		Sulfate SO_4^{2-}	Nitrate NO_3^-
Lithium Li^+	Magnesium Mg^{2+}	All Group 6 elements, including:	All Group 7 elements, including:
Sodium Na^+	Calcium Ca^{2+}	Oxide O^{2-}	Fluoride F^-
Potassium K^+		Sulfide S^{2-}	Chloride Cl^-
			Bromide Br^-
			Iodide I^-

9) Any of the positive ions above can combine with any of the negative ions to form an ionic compound.

10) Only elements at opposite sides of the periodic table will form ionic compounds, e.g. Na and Cl, where one of them becomes a positive ion and one becomes a negative ion.

> Remember, the + and − charges we talk about, e.g. Na^+ for sodium, just tell you what type of ion the atom WILL FORM in a chemical reaction. In sodium metal there are only neutral sodium atoms, Na. The Na^+ ions will only appear if the sodium metal reacts with something like water or chlorine.

Look at Charges to Work Out the Formula of an Ionic Compound

1) Ionic compounds are made up of a positively charged part and a negatively charged part.

2) The overall charge of any compound is zero.

3) So all the negative charges in the compound must balance all the positive charges.

4) You can use the charges on the individual ions present to work out the formula for the ionic compound:

> Sodium chloride contains Na^+ (+1) and Cl^- (−1) ions.
>
> (+1) + (−1) = 0. The charges are balanced with one of each ion, so the formula for sodium chloride = NaCl

> Magnesium chloride contains Mg^{2+} (+2) and Cl^- (−1) ions.
>
> Because a chloride ion only has a 1− charge we will need two of them to balance out the 2+ charge of a magnesium ion. This gives us the formula $MgCl_2$.

The formula for exam success = revision...

Remember, the + and − charges only appear when an element reacts with something. They're neutral otherwise.

Electronic Structure of Ions

If you fancy yourself as a bit of an artist then you're in luck. This page is full of lovely drawings of <u>electronic structures</u> that should put a smile on your face.

Show the Electronic Structure of Simple Ions With Diagrams

A useful way of representing ions is by <u>drawing</u> out their electronic structure. Just use a big <u>square bracket</u> and a + or − to show the charge. A few <u>ions</u> and the <u>ionic compounds</u> they form are shown below.

Sodium Chloride

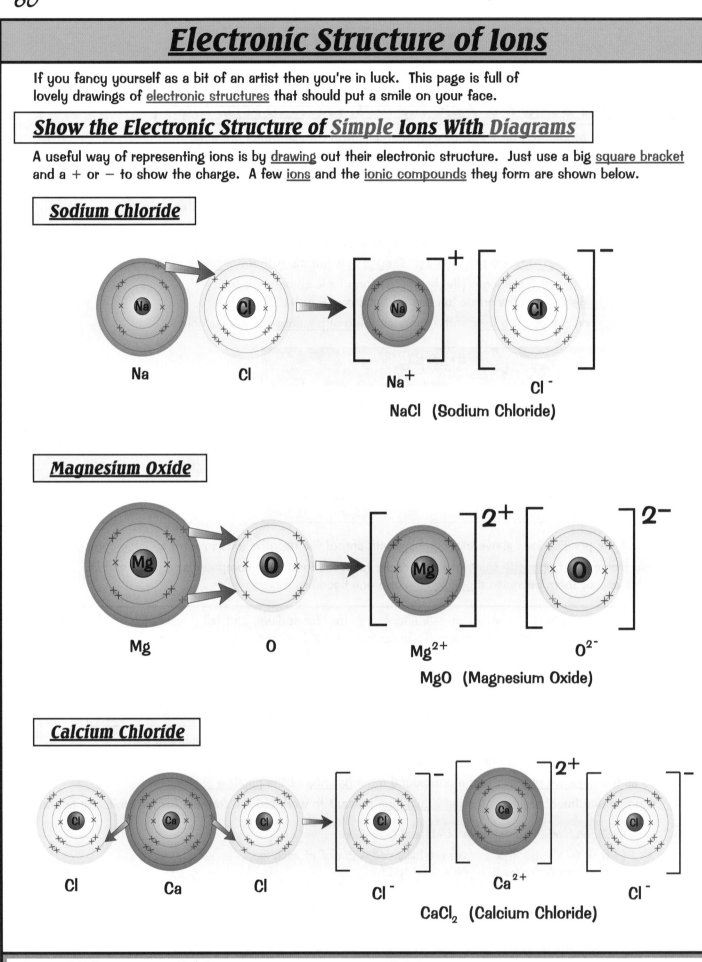

Na Cl Na$^+$ Cl$^-$

NaCl (Sodium Chloride)

Magnesium Oxide

Mg O Mg^{2+} O^{2-}

MgO (Magnesium Oxide)

Calcium Chloride

Cl Ca Cl Cl$^-$ Ca^{2+} Cl$^-$

CaCl$_2$ (Calcium Chloride)

Any old ion, any old ion — any, any, any old ion...

3 ionic compounds, 3 drawings, 1 exam hall. It's like the start of a bad Game Show. Whether or not you're able to produce some lovely drawings of these bad boys all comes down to how well you've understood <u>ionic bonding</u>. (So if you're struggling, try reading the last few pages again — I know I had to.)

Group 1 — The Alkali Metals

Alkali metals are all members of the same group — Group 1. That means they'll all have similar properties.

Group 1 Metals are Known as the 'Alkali Metals'

1) Group 1 metals include lithium, sodium and potassium.

2) They all have ONE outer shell electron. This makes them very reactive and gives them all similar properties.

3) When the alkali metals react they all form similar compounds (see below).

4) The alkali metals are shiny when freshly cut, but quickly react with the oxygen in moist air and tarnish.

As you go DOWN Group 1, the alkali metals:
1) become MORE REACTIVE (see below)

...because the outer electron is more easily lost, because it's further from the nucleus.

2) have a HIGHER DENSITY

...because the atoms have more mass.

3) have a LOWER MELTING POINT

4) have a LOWER BOILING POINT

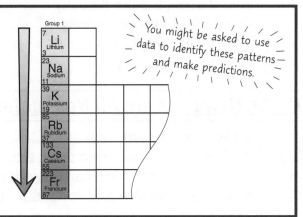

You might be asked to use data to identify these patterns and make predictions.

Reaction with Cold Water Produces Hydrogen Gas

1) When lithium, sodium or potassium are put in water, they react very vigorously.

2) They move around the surface, fizzing furiously.

3) They produce hydrogen. Potassium gets hot enough to ignite it. If it hasn't already been ignited by the reaction, a lighted splint will indicate hydrogen by producing the notorious "squeaky pop" as it ignites.

4) The reaction makes an alkaline solution — this is why Group 1 is known as the alkali metals.

5) A hydroxide of the metal forms, e.g. sodium hydroxide (NaOH) or potassium hydroxide (KOH).

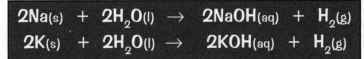

$$2Na_{(s)} + 2H_2O_{(l)} \rightarrow 2NaOH_{(aq)} + H_{2(g)}$$
$$2K_{(s)} + 2H_2O_{(l)} \rightarrow 2KOH_{(aq)} + H_{2(g)}$$

6) This experiment shows the relative reactivities of the alkali metals. The more violent the reaction, the more reactive the alkali metal is.

The Alkali Metals Form Ionic Compounds with Non-Metals

1) They are keen to lose their one outer electron to form a 1+ ion.

2) They are so keen to lose the outer electron there's no way they'd consider sharing, so covalent bonding is out of the question.

3) So they always form ionic bonds — and they produce white compounds that dissolve in water to form colourless solutions.

Notorious Squeaky Pop — a.k.a. the Justin Timberlake test...

Alkali metals are ace. They're so reactive you have to store them in oil — because otherwise they'd react with the water vapour in the air. AND they fizz in water and explode and everything. Cool.

Group 7 — The Halogens

There are trends in Group 7 as well. But some of the trends are kind of the opposite of the Group 1 trends.

Reactivity Decreases Down the Group

As you go DOWN Group 7, the HALOGENS have the following properties:

1) LESS REACTIVE
   ...because it's harder to gain an extra electron, because the outer shell's further from the nucleus.
2) HIGHER MELTING POINT
3) HIGHER BOILING POINT

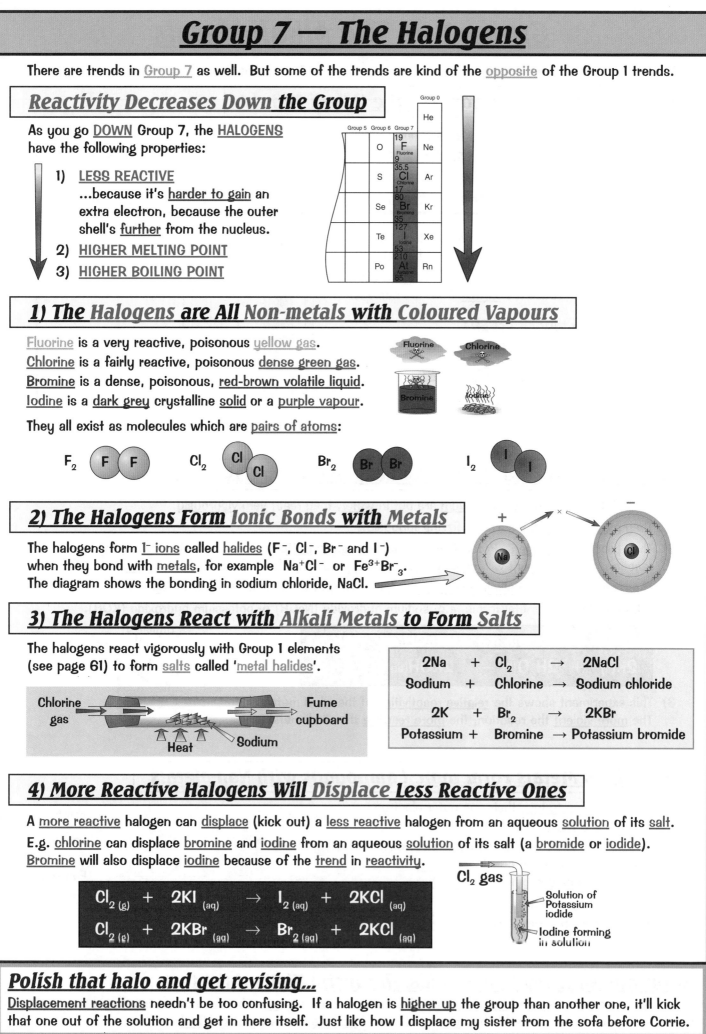

1) The Halogens are All Non-metals with Coloured Vapours

Fluorine is a very reactive, poisonous yellow gas.
Chlorine is a fairly reactive, poisonous dense green gas.
Bromine is a dense, poisonous, red-brown volatile liquid.
Iodine is a dark grey crystalline solid or a purple vapour.

They all exist as molecules which are pairs of atoms:

F_2 F F Cl_2 Cl Cl Br_2 Br Br I_2 I I

2) The Halogens Form Ionic Bonds with Metals

The halogens form 1− ions called halides (F^-, Cl^-, Br^- and I^-) when they bond with metals, for example Na^+Cl^- or $Fe^{3+}Br^-_3$. The diagram shows the bonding in sodium chloride, NaCl.

3) The Halogens React with Alkali Metals to Form Salts

The halogens react vigorously with Group 1 elements (see page 61) to form salts called 'metal halides'.

Chlorine gas   Fume cupboard
Heat   Sodium

$$2Na + Cl_2 \rightarrow 2NaCl$$
Sodium + Chlorine → Sodium chloride
$$2K + Br_2 \rightarrow 2KBr$$
Potassium + Bromine → Potassium bromide

4) More Reactive Halogens Will Displace Less Reactive Ones

A more reactive halogen can displace (kick out) a less reactive halogen from an aqueous solution of its salt.

E.g. chlorine can displace bromine and iodine from an aqueous solution of its salt (a bromide or iodide). Bromine will also displace iodine because of the trend in reactivity.

$$Cl_{2\,(g)} + 2KI_{(aq)} \rightarrow I_{2\,(aq)} + 2KCl_{(aq)}$$
$$Cl_{2\,(g)} + 2KBr_{(aq)} \rightarrow Br_{2\,(aq)} + 2KCl_{(aq)}$$

Cl_2 gas
Solution of Potassium iodide
Iodine forming in solution

Polish that halo and get revising...

Displacement reactions needn't be too confusing. If a halogen is higher up the group than another one, it'll kick that one out of the solution and get in there itself. Just like how I displace my sister from the sofa before Corrie.

Group 0 — The Noble Gases

The <u>noble gases</u> — stuffed full of every honourable virtue. They <u>don't react</u> with very much and you can't even see them — making them, well, a bit <u>dull</u> really.

Group 0 Elements are All <u>Inert, Colourless Gases</u>

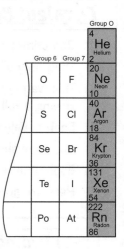

1) Group 0 elements are called the <u>noble gases</u> and include the elements <u>helium</u>, <u>neon</u> and <u>argon</u> (plus a few others).

2) All elements in Group 0 are <u>colourless gases</u> at room temperature.

3) They are also more or less <u>inert</u> — this means they <u>don't react</u> with much at all. The reason for this is that they have a <u>full outer shell</u>. This means they're <u>not</u> desperate to <u>give up</u> or <u>gain</u> electrons.

4) As the noble gases are inert they're <u>non-flammable</u> — they won't set on fire.

5) These properties make the gases <u>hard to observe</u> — it took a long time for them to be discovered.

6) The gases were found when chemists noticed that the <u>density</u> of <u>nitrogen</u> made in chemical <u>reactions</u> was <u>different</u> to the density of nitrogen taken from the <u>air</u>.

7) They hypothesised that the nitrogen obtained from air must have <u>other gases mixed in with it</u>.

8) Scientists gradually discovered the different noble gases through a series of <u>experiments</u>, including the <u>fractional distillation</u> of <u>air</u> (see page 50).

The Noble Gases have <u>Many Everyday Uses...</u>

<u>Argon</u> is used to provide an <u>inert atmosphere</u> in <u>filament lamps</u> (light bulbs). As the argon is <u>non-flammable</u> it stops the very hot filament from <u>burning away</u>. It can also be used to protect metals that are being <u>welded</u>. The inert atmosphere stops the hot metal reacting with <u>oxygen</u>.

<u>Helium</u> is used in <u>airships</u> and <u>party balloons</u>. Helium has a <u>lower density</u> than air — so it makes balloons <u>float</u>.

There are <u>Patterns</u> <u>in</u> the <u>Properties</u> of the Noble Gases

The <u>boiling points</u> and <u>densities</u> of the noble gases <u>increase</u> as you move <u>down</u> the group.

Noble Gas	Boiling Point (°C)	Density (g/cm³)
helium	-269	0.0002
neon	-246	0.0009
argon	-186	0.0018
krypton	-153	0.0037
xenon	-108	0.0059
radon	-62	0.0097

They don't react — that's Noble De use to us chemists...

The noble gases seem more dull than a brown bread sandwich but they're not so bad. They'd be pretty good in a game of <u>hide and seek</u> for a start. And what would <u>balloon sellers</u> be without them? Deflated — that's what.

Covalent Bonding

Some elements form ionic bonds (see page 58) but others form strong covalent bonds. This is where atoms share electrons with each other so that they've got full outer shells.

Covalent Bonds — Sharing Electrons

1) Sometimes atoms prefer to make covalent bonds by sharing electrons with other atoms.

2) They only share electrons in their outer shells (highest energy levels).

3) This way both atoms feel that they have a full outer shell, and that makes them happy. Having a full outer shell gives them the electronic structure of a noble gas.

4) Each covalent bond provides one extra shared electron for each atom.

5) So, a covalent bond is a shared pair of electrons.

6) Each atom involved has to make enough covalent bonds to fill up its outer shell.

7) Here are some important examples:

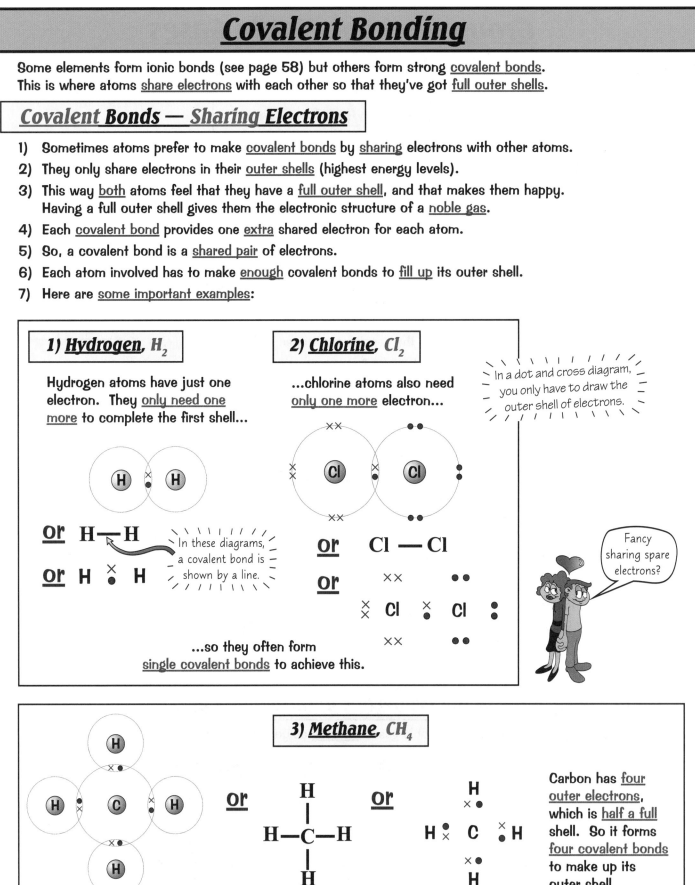

1) Hydrogen, H₂

Hydrogen atoms have just one electron. They only need one more to complete the first shell...

or H—H

In these diagrams, a covalent bond is shown by a line.

or H × ● H

2) Chlorine, Cl₂

...chlorine atoms also need only one more electron...

In a dot and cross diagram, you only have to draw the outer shell of electrons.

or Cl — Cl

or

...so they often form single covalent bonds to achieve this.

Fancy sharing spare electrons?

3) Methane, CH₄

or

H
|
H—C—H
|
H

or

Carbon has four outer electrons, which is half a full shell. So it forms four covalent bonds to make up its outer shell.

Covalent bonding — it's good to share...

There's another page of covalent bonding diagrams yet to come, but make sure you can draw the diagrams for the covalent compounds on this page first. When you've drawn a dot and cross diagram, it's a really good idea to count up the number of electrons, just to double check you've definitely got a full outer shell.

More Covalent Bonding

You lucky thing. There are five more examples of covalent bonding on this page — and for each compound there are three possible <u>diagrams</u>. I make that fifteen diagrams in total... and just a smattering of words. So, this page is a breeze compared to others out there.

4) <u>Hydrogen Chloride</u>, HCl

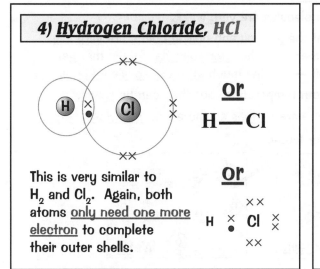

or

H — Cl

or

This is very similar to H_2 and Cl_2. Again, both atoms <u>only need one more electron</u> to complete their outer shells.

5) <u>Ammonia</u>, NH_3

Nitrogen has <u>five</u> outer electrons...

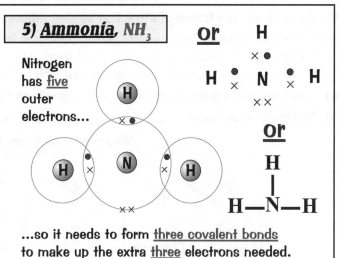

...so it needs to form <u>three covalent bonds</u> to make up the extra <u>three</u> electrons needed.

Remember — it's only the outer shells that share electrons with each other.

6) <u>Water</u>, H_2O

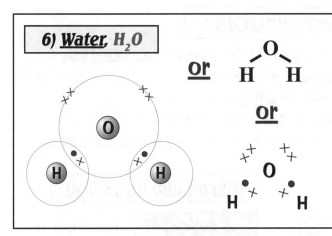

or

or

<u>Oxygen</u> atoms have <u>six</u> outer electrons. They sometimes form <u>ionic</u> bonds by <u>taking</u> two electrons to complete their outer shell. However they'll also cheerfully form <u>covalent bonds</u> and <u>share</u> two electrons instead. In <u>water molecules</u>, the oxygen <u>shares</u> electrons with the two H atoms.

7) <u>Oxygen</u>, O_2

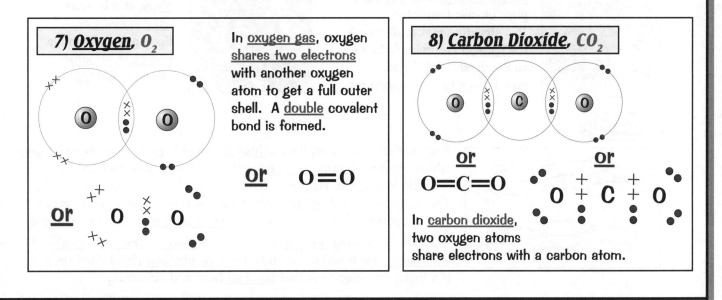

In <u>oxygen gas</u>, oxygen <u>shares two electrons</u> with another oxygen atom to get a full outer shell. A <u>double</u> covalent bond is formed.

or O = O

8) <u>Carbon Dioxide</u>, CO_2

or
O = C = O

or

In <u>carbon dioxide</u>, two oxygen atoms share electrons with a carbon atom.

The name's Bond, Covalent Bond...

Every atom wants a full outer shell, and they can get that either by becoming an <u>ion</u> (see page 58) or by <u>sharing electrons</u>. Once you understand that, you should be able to apply it to any example. Piece of cake.

Covalent Substances: Two Kinds

Substances with <u>covalent bonds</u> (electron sharing) can either be <u>simple molecules</u> or <u>giant structures</u>.

Simple Molecular Substances

1) The atoms form <u>very strong</u> covalent bonds to form <u>small</u> molecules of several atoms.

2) By contrast, the forces of attraction <u>between</u> these molecules are <u>very weak</u>.

3) The result of these feeble <u>intermolecular forces</u> is that the <u>melting</u> and <u>boiling points</u> are <u>very low</u>, because the molecules are <u>easily parted</u> from each other. It's the <u>intermolecular forces</u> that get <u>broken</u> when simple molecular substances melt or boil — <u>not</u> the much <u>stronger covalent bonds</u>.

4) Most molecular substances are <u>gases or liquids</u> at room temperature, but they can be <u>solids</u>.

5) Molecular substances <u>don't conduct electricity</u> — there are <u>no ions</u> so there's <u>no electrical charge</u>.

Very weak intermolecular forces

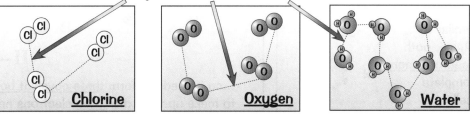

Chlorine   Oxygen   Water

Giant Covalent Structures Are Macromolecules

1) These are similar to giant ionic structures (lattices) <u>except</u> that there are <u>no charged ions</u>.

2) <u>All</u> the atoms are <u>bonded</u> to <u>each other</u> by <u>strong</u> covalent bonds.

3) This means that they have <u>very high</u> melting and boiling points.

4) They <u>don't conduct electricity</u> — not even when <u>molten</u> (except for graphite).

5) The <u>main examples</u> are <u>diamond</u> and <u>graphite</u>, which are both made only from <u>carbon atoms</u>, and <u>silicon dioxide</u> (silica).

Diamond

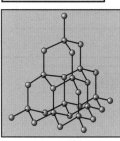

Each carbon atom forms <u>four</u> <u>covalent bonds</u> in a <u>very rigid</u> giant covalent structure. This structure makes diamond the <u>hardest</u> natural substance, so it's used for drill tips. And it's <u>pretty</u> and <u>sparkly</u> too.

Silicon Dioxide (Silica)

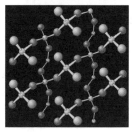

Sometimes called <u>silica</u>, this is what <u>sand</u> is made of. Each grain of sand is <u>one giant structure</u> of silicon and oxygen.

Graphite

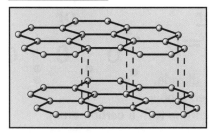

Each carbon atom only forms <u>three covalent bonds</u>. This creates <u>layers</u> which are free to <u>slide over each other</u>, like a pack of cards — so graphite is <u>soft</u> and <u>slippery</u>. The layers are held together so loosely that they can be <u>rubbed off</u> onto paper — that's how a <u>pencil</u> works. This is because there are <u>weak intermolecular forces</u> between the layers.

Graphite is the only <u>non-metal</u> which is a <u>good conductor of heat and electricity</u>. Each carbon atom has one <u>delocalised</u> (free) electron and it's these free electrons that <u>conduct</u> heat and electricity.

Carbon is a girl's best friend...

The <u>two different types</u> of covalent substance are very different. Make sure you don't get them muddled up.

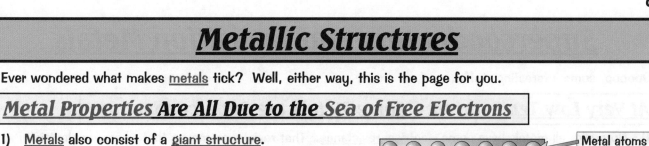

Ever wondered what makes <u>metals</u> tick? Well, either way, this is the page for you.

<u>Metal Properties</u> <u>Are All Due to the</u> <u>Sea of Free Electrons</u>

1) <u>Metals</u> also consist of a <u>giant structure</u>.

2) <u>Metallic bonds</u> involve the all-important '<u>free electrons</u>' which produce <u>all</u> the properties of metals. These delocalised (free) electrons come from the <u>outer shell</u> of <u>every</u> metal atom in the structure.

3) These electrons are <u>free to move</u> through the whole structure and so metals are good conductors of <u>heat and electricity</u>.

4) These electrons also <u>hold</u> the <u>atoms</u> together in a <u>regular</u> structure. There are strong forces of <u>electrostatic attraction</u> between the <u>positive metal ions</u> and the <u>negative electrons</u>.

5) They also allow the layers of atoms to <u>slide</u> over each other, allowing metals to be <u>bent</u> and <u>shaped</u>.

<u>Alloys</u> <u>are Harder</u> <u>Than Pure Metals</u>

1) <u>Pure metals</u> often aren't quite right for certain jobs. So scientists <u>mix two or more metals together</u> — creating an <u>alloy</u> with the properties they want.

2) Different elements have <u>different sized atoms</u>. So when another metal is mixed with a pure metal, the new metal atoms will <u>distort</u> the layers of metal atoms, making it more difficult for them to slide over each other. So alloys are <u>harder</u>.

<u>Identifying</u> <u>the Structure of a Substance</u> <u>by Its Properties</u>

You can <u>identify</u> most substances just by the way they <u>behave</u> as either:

- <u>giant ionic</u>,
- <u>simple molecular</u>,
- <u>giant covalent</u>,
- or <u>giant metallic</u>.

<u>Example</u>: Four substances were tested for various properties with the following results:

Substance	Melting point (°C)	Boiling point (°C)	Good electrical conductor?
A	–218.4	–182.96	No
B	1535	2750	Yes
C	1410	2355	No
D	801	1413	When molten

Identify the structure of each substance. (Answers on page 140.)

<u>A few free electrons and my knees have gone all bendy...</u>

If you know the <u>properties</u> of a substance then you can work out its <u>structure</u>. And explain <u>why</u>, of course.

Superconductors and Transition Metals

Oooooo, some interesting stuff...

At Very Low Temperatures, Some Metals are Superconductors

1) Normally, all metals have some electrical resistance. That resistance means that whenever electricity flows through them, they heat up, and some of the electrical energy is wasted as heat.

2) If you make some metals cold enough, though, their resistance disappears completely. The metal becomes a superconductor.

3) Without any resistance, none of the electrical energy is turned into heat, so none of it's wasted.

4) That means you could start a current flowing through a superconducting circuit, take out the battery, and the current would carry on flowing forever.

So What's the Catch...

1) Using superconducting wires you can make:

   a) Power cables that transmit electricity without any loss of power (loss-free power transmission).

   b) Really strong electromagnets that don't need a constant power source.

   c) Electronic circuits that work really fast, because there's no resistance to slow them down.

2) But they need to be REALLY COLD. Metals only start superconducting at less than −265 °C! Getting things that cold is very hard, and very expensive, which limits the use of superconductors.

3) Scientists are trying to develop room temperature superconductors now. So far, they've managed to get some weird metal oxide things to superconduct at about −135 °C, which is a much cheaper temperature to get down to. But ideally they need to develop superconductors that still work at 20 °C.

Metals in the Middle of the Periodic Table are Transition Metals

A lot of everyday metals are transition metals (e.g. copper, iron, zinc, gold, silver, platinum) — but there are lots of others as well. Transition metals have typical 'metallic' properties.

If you get asked about a transition metal you've never heard of — don't panic. These 'new' transition metals follow all the properties you've already learnt for the others.

These are the transition metals

| | Sc | Ti | V | Cr | Mn | Fe | Co | Ni | Cu | Zn | |

Transition metals and their compounds make good catalysts

1) Iron is the catalyst used in the Haber process for making ammonia.

2) Nickel is useful for the hydrogenation of alkenes (e.g. to make margarine).

Transition metals often have more than one ion, e.g. Fe^{2+}, Fe^{3+}

Two other examples are copper: Cu^+ and Cu^{2+} and chromium: Cr^{2+} and Cr^{3+}.

The compounds are very colourful

The compounds of transition elements are colourful due to the transition metal ion they contain. E.g. Iron(II) compounds are usually light green, iron(III) compounds are orange/brown (e.g. rust) and copper compounds are often blue.

Mendeleev and his amazing technicoloured periodic table...

Superconducting magnets are used in magnetic resonance image (MRI) scanners in hospitals. That way, the huge magnetic fields they need can be generated without using up a load of electricity. Great stuff...

Thermal Decomposition and Precipitation

It's your lucky day. There is not one, but __TWO__ exciting reactions on this page. Off yer go...

1) Thermal Decomposition — Breaking Down with Heat

1) Thermal decomposition is when a substance breaks down into at least two other substances when heated.

2) Transition metal carbonates break down on heating. Transition metal carbonates are things like copper(II) carbonate ($CuCO_3$), iron(II) carbonate ($FeCO_3$), zinc carbonate ($ZnCO_3$) and manganese carbonate ($MnCO_3$), i.e. they've all got a CO_3 bit in them.

3) They break down into a metal oxide (e.g. copper oxide, CuO) and carbon dioxide. This usually results in a colour change.

> The reactions for the thermal decomposition of:
> (i) iron(II) carbonate to iron oxide (FeO),
> (ii) manganese carbonate to manganese oxide (MnO),
> (iii) zinc carbonate to zinc oxide (ZnO),
> are the same — although the colours are different.

__EXAMPLE:__ The thermal decomposition of copper(II) carbonate.

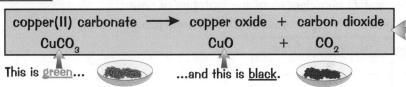

copper(II) carbonate ⟶ copper oxide + carbon dioxide
$CuCO_3$ ⟶ CuO + CO_2

This is green... ...and this is black.

4) You can easily check that the gas given off is carbon dioxide by bubbling it through limewater. If carbon dioxide is present the limewater turns milky.

2) Precipitation — A Solid Forms in Solution

1) A precipitation reaction is where two solutions react and an insoluble solid forms in the solution.

2) The solid is said to 'precipitate out' and, confusingly, the solid is also called 'a precipitate'.

3) Some soluble transition metal compounds react with sodium hydroxide to form an insoluble hydroxide, which then precipitates out. Here are some examples:

$CuSO_4$ + $2NaOH$ ⟶ $Cu(OH)_2$ + Na_2SO_4
copper(II) sulfate + sodium hydroxide ⟶ copper(II) hydroxide + sodium sulfate

$FeSO_4$ + $2NaOH$ ⟶ $Fe(OH)_2$ + Na_2SO_4
iron(II) sulfate + sodium hydroxide ⟶ iron(II) hydroxide + sodium sulfate

$Fe_2(SO_4)_3$ + $6NaOH$ ⟶ $2Fe(OH)_3$ + $3Na_2SO_4$
iron(III) sulfate + sodium hydroxide ⟶ iron(III) hydroxide + sodium sulfate

4) You can also write the above equations in terms of ions, for example:

$$Cu^{2+} + 2OH^- \longrightarrow Cu(OH)_2$$

Use Precipitation to Test for Transition Metal Ions

1) Some insoluble transition metal hydroxides have distinctive colours.

2) You can use this fact to test which transition metal ions a solution contains.

> Copper(II) hydroxide is a blue solid.
> Iron(II) hydroxide is a grey/green solid.
> Iron(III) hydroxide is an orange/brown solid.

3) For example, if you add sodium hydroxide to an unknown soluble salt, and an orange/brown precipitate forms, you know you've got iron(III) hydroxide and so have Fe^{3+} ions in the solution.

My duffel coat's worn out — thermal decomposition...

__Bad news:__ this page is packed full of chemistry. And it's not easy either. __Good news:__ it's the end of another section. But before you get a well earned biccie, have a go at the questions on the next page. Cheers, duck.

Revision Summary for Section Six

And just when you realise that you can't take any more Chemistry, it's the end of the section. Phew. Which means it's time for some more questions. You probably want to rip the page out and eat it — but they're the best way of testing that you've learned everything in this topic. If you can't answer any of them, look back in the book. If you can't do all this now, you won't be able to in the exam either.

1) What size groups did Döbereiner organise the elements into? What were these groups called?

2) Give two reasons why Newlands' Octaves were criticised.

3) Why did Mendeleev leave gaps in his Table of Elements?

4) How are the group number and the number of electrons in the outer shell related?

5) What is shielding?

6) Define the term isotope.

7)* The table below gives the masses and relative abundances of the isotopes of neon:
   Calculate the relative atomic mass of neon. Give your answer to 2 decimal places.

relative mass of isotope	relative abundance
20	91%
22	9%

8) Describe the structure of a crystal of sodium chloride.

9) List the main properties of ionic compounds.

10) What type of ion do elements from the following groups form?
    a) Group 1   b) Group 7

11) Use information from the periodic table to help you work out the formulas of these ionic compounds:
    a) potassium chloride   b) calcium chloride

12) Which group are the alkali metals?

13) As you go down the group of alkali metals, do they become more or less reactive?

14) Give details of the reactions of the alkali metals with water.

15) Explain why Group 7 elements get less reactive as you go down the group from fluorine to iodine.

16) Write down the balanced equation for the displacement of bromine from potassium bromide by chlorine.

17) Will the following reactions occur: a) iodine with lithium chloride, b) chlorine with lithium bromide?

18) Explain why the noble gases are inert.

19) Name two noble gases and state a use for each.

20) Sketch dot and cross diagrams showing the bonding in molecules of:
    a) hydrogen,   b) hydrogen chloride,   c) water,   d) ammonia

21) What are the two types of covalent substance? Give three examples of each.

22) List three properties of metals and explain how metallic bonding causes these properties.

23) Explain why alloys are harder than pure metals.

24)*Identify the structure of each of the substances in the table:

Substance	Melting point (°C)	Electrical conductivity	Hardness [scale of 0 – 10 (10 being diamond)]
A	3410	Very high	7.5
B	2072	Zero	9
C	605	Zero in solid form High when molten	Low

25) What is a superconductor? Describe some useful applications of superconductors.

26) Name six transition metals, and give uses for two of them.

27) What are thermal decomposition reactions?

28) What type of reaction between two liquids results in the formation of a solid?
    What are these solid products called?

* Answers on page 140.

Relative Formula Mass

The biggest trouble with <u>relative atomic mass</u> and <u>relative formula mass</u> is that they <u>sound</u> so blood-curdling. Take a few deep breaths, and just enjoy, as the mists slowly clear...

Relative Atomic Mass, A_r — Easy Peasy

1) You'll probably remember about <u>relative atomic mass</u> A_r from page 57.

2) The A_r for each element is the <u>same</u> as its <u>mass number</u> (see page 11).

3) So, you can find it easily by looking at the <u>periodic table</u> — the <u>bigger number</u> for each element is the <u>relative atomic mass</u>. For example:

<u>Relative atomic mass</u>

$$^{4}_{2}\text{He} \qquad ^{12}_{6}\text{C}$$

Relative Formula Mass, M_r — Also Easy Peasy

If you have a compound like $MgCl_2$ then it has a <u>relative formula mass</u>, M_r, which is just all the relative atomic masses <u>added together</u>.
For $MgCl_2$ it would be:

The relative atomic mass of chlorine is multiplied by 2 because there are two chlorine atoms.

$$\text{MgCl}_2$$

$$24 \quad + \quad (35.5 \times 2) \quad = \quad 95$$

So M_r for $MgCl_2$ is simply <u>95</u>.

You can easily get A_r for any element from the periodic table (see inside front cover), but in a lot of questions they give you them anyway. And that's all it is. A big fancy name like <u>relative formula mass</u> and all it means is "<u>add up all the relative atomic masses</u>". What a swizz, eh?

"ONE MOLE" of a Substance is Equal to its M_r in Grams

The <u>relative formula mass</u> (A_r or M_r) of a substance <u>in grams</u> is known as <u>one mole</u> of that substance.

<u>Examples</u>:
Iron has an A_r of 56.   So one mole of iron weighs exactly 56 g
Nitrogen gas, N_2, has an M_r of 28 (2 × 14).   So one mole of N_2 weighs exactly 28 g

You can convert between moles and grams using this formula:

$$\text{NUMBER OF MOLES} = \frac{\text{Mass in g (of element or compound)}}{M_r \text{ (of element or compound)}}$$

<u>Example</u>: How many moles are there in 42 g of carbon?
<u>Answer</u>: No. of moles = Mass (g) / M_r = 42/12 = <u>3.5 moles</u>   Easy Peasy

Numbers? — and you thought you were doing chemistry...

If you want to crack <u>relative atomic mass</u> and <u>relative formula mass</u>, having a go at these should help:
1) Use the periodic table to find the relative atomic mass of these elements: Cu, K, Kr, Cl
2) Find the relative formula mass of: NaOH, Fe_2O_3, C_6H_{14}, $Mg(NO_3)_2$ Answers on page 140.

Two Formula Mass Calculations

Although relative atomic mass and relative formula mass are <u>easy enough</u>, it can get just a tad <u>trickier</u> when you start getting into other calculations which use them. It depends on how good your maths is basically, because it's all to do with ratios and percentages.

Calculating % Mass of an Element in a Compound

This is actually dead easy — so long as you've learnt this formula:

$$\text{Percentage mass OF AN ELEMENT IN A COMPOUND} = \frac{A_r \times \text{No. of atoms (of that element)}}{M_r \text{ (of whole compound)}} \times 100$$

<u>EXAMPLE:</u> Find the percentage mass of sodium in sodium carbonate, Na_2CO_3.

<u>ANSWER:</u>

A_r of sodium = 23, A_r of carbon = 12, A_r of oxygen = 16

M_r of Na_2CO_3 = $(2 \times 23) + 12 + (3 \times 16) = 106$

Now use the formula:

$$\underline{\text{Percentage mass}} = \frac{A_r \times n}{M_r} \times 100 = \frac{23 \times 2}{106} \times 100 = 43.4\%$$

And there you have it. Sodium makes up <u>43.4%</u> of the mass of sodium carbonate.

Finding the Empirical Formula (from Masses or Percentages)

This also sounds a lot worse than it really is. Try this for an easy peasy <u>stepwise method</u>:

1) <u>List all the elements</u> in the compound (there's usually only two or three!)

2) <u>Underneath them</u>, write their <u>experimental masses or percentages</u>.

3) <u>Divide</u> each mass or percentage <u>by the A_r</u> for that particular element.

4) Turn the numbers you get into <u>a nice simple ratio</u> by multiplying and/or dividing them by well-chosen numbers.

5) Get the ratio in its <u>simplest form</u>, and that tells you the <u>empirical formula</u> of the compound.

Example: Find the empirical formula of the iron oxide produced when 44.8 g of iron react with 19.2 g of oxygen. (A_r for iron = 56, A_r for oxygen = 16)

Method:

1) <u>List the two elements:</u>   Fe   O

2) Write in the <u>experimental masses</u>:   44.8   19.2

3) <u>Divide by the A_r</u> for each element: $\frac{44.8}{56} = 0.8$ $\frac{19.2}{16} = 1.2$

4) Multiply by 10...   8   12

     ...then divide by 4:   2   3

5) So the <u>simplest formula</u> is 2 atoms of Fe to 3 atoms of O, i.e. <u>Fe_2O_3</u>. And that's it done.

> This <u>empirical method</u> (i.e. based on <u>experiment</u>) is the <u>only way</u> of finding out the formula of a compound. Rust is iron oxide, sure, but is it FeO, or Fe_2O_3? Only an experiment to determine the empirical formula will tell you for certain.

With this empirical formula I can rule the world! — mwa ha ha ha...

With calculations it's a case of good old fashioned practice makes perfect. Try these: Answers on page 140.

1) Find the percentage mass of oxygen in each of these: a) Fe_2O_3 b) H_2O c) $CaCO_3$ d) H_2SO_4.

2) Find the empirical formula of the compound formed from 2.4 g of carbon and 0.8 g of hydrogen.

Section Seven — Equations and Calculations

Molar Volume and Conservation of Mass

This page has the answers to some pretty exciting chemistry-related questions — exactly how much space does a <u>mole of gas</u> take up... and what does happen to <u>mass</u> in a chemical reaction... ? Hot stuff.

One Mole of Gas Occupies a Volume of 24 dm³

Remember moles from page 71? Well here's a wee bit more about them:

Remember dm³ is just a fancy way of writing 'litre', so 1 dm³ = 1000 cm³

<u>One mole</u> of <u>any gas</u> always occupies <u>24 dm³</u> (= 24 000 cm³) at room temperature and pressure (RTP = 25 °C and 1 atmosphere)

Example 1: What's the volume of 4.5 moles of chlorine at RTP?

Answer: 1 mole = 24 dm³, so 4.5 moles = 4.5 × 24 dm³ = <u>108 dm³</u>

Example 2: How many moles are there in 8280 cm³ of hydrogen gas at RTP?

Answer: Number of moles = $\dfrac{\text{Volume of gas}}{\text{Volume of 1 mole}}$ = $\dfrac{8.28}{24}$ = <u>0.345 moles</u>

Don't forget to convert from cm³ to dm³.

Volume / Moles × 24

In a Chemical Reaction, Mass is Always Conserved

1) During a chemical reaction <u>no atoms are destroyed</u> and <u>no atoms are created</u>.

2) This means there are the <u>same number and types of atoms</u> on each side of a reaction equation.

3) Because of this no mass is lost or gained — we say that mass is <u>conserved</u> during a reaction.

<u>Example</u>: 2Li + F$_2$ → 2LiF

<u>Method</u>: There are <u>2</u> lithium atoms and <u>2</u> fluorine atoms on <u>each side</u> of the equation.

4) By adding up the relative formula masses on each side of the equation you can see that mass is conserved.

<u>Example</u>: 2Li + F$_2$ → 2LiF

<u>Method</u>: (2 × 7) + (2 × 19) → 2 × 26

14 + 38 → 52

52 → 52 So, <u>mass is conserved</u> in this equation.

5) You can use simple <u>ratios</u> to calculate the reacting masses in a reaction.

<u>Example</u>: In the reaction, 2Li + F$_2$ → 2LiF, 14 g of lithium will react with 38 g of fluorine.

<u>Method</u>: The only product that's formed is lithium fluoride, so 14 + 38 = 52 g will be produced.

The masses for this reaction will always be in the same proportions as this.

Multiplying or dividing these masses by the same number gives you other sets of reacting masses.

Element / compound in reaction	Lithium	Fluorine	Lithium fluoride
Original reacting masses	14 g	38 g	52 g
Reacting masses set 2	14 ÷ 2 = 7 g	38 ÷ 2 = 19 g	52 ÷ 2 = 26 g
Reacting masses set 3	14 × 1.5 = 21 g	38 × 1.5 = 57 g	52 × 1.5 = 78 g

Phew, Chemistry — scary stuff sometimes, innit...

Q: How much space does one mole of gas take up at RTP? A: 24 dm³ and not a whisker more.
Q: What happens to total mass in a chemical reaction? A: Nothing. It just stays the same. Yawn.

Calculating Masses in Reactions

Sorry — another page of mean looking calculations, but chill out, little trembling one — just relax and enjoy.

The Three Important Steps — Not to Be Missed...

(Miss one out and it'll all go horribly wrong, believe me.)

1) <u>Write out</u> the balanced <u>equation</u>
2) <u>Work out</u> M_r — just for the <u>two bits you want</u>
3) Apply the rule: <u>Divide to get one, then multiply to get all</u>
(But you have to apply this first to the substance they give you information about, and then the other one!)

Don't worry — these steps should all make sense when you look at the example below.

<u>Example</u>: What mass of magnesium oxide is produced when 60 g of magnesium is burned in air?

<u>Answer</u>:

1) Write out the <u>balanced equation</u>:

$$2Mg + O_2 \rightarrow 2MgO$$

2) Work out the <u>relative formula masses</u>:

(don't do the oxygen — we don't need it)

$$2 \times 24 \qquad \rightarrow \qquad 2 \times (24+16)$$
$$48 \qquad \rightarrow \qquad 80$$

3) Apply the rule: <u>Divide to get one, then multiply to get all</u>:

The two numbers, 48 and 80, tell us that <u>48 g of Mg react to give 80 g of MgO</u>.
Here's the tricky bit. You've now got to be able to write this down:

> 48 g of Mg ......reacts to give.....80 g of MgO
>
> 1 g of Mg ......reacts to give.....
>
> 60 g of Mg .....reacts to give......

<u>The big clue</u> is that in the question they've said we want to burn "<u>60 g of magnesium</u>",
i.e. they've told us how much <u>magnesium</u> to have, and that's how you know to write down the
<u>left-hand side</u> of it first, because:

We'll first need to ÷ by 48 to get 1 g of Mg
and then need to × by 60 to get 60 g of Mg.

<u>Then</u> you can work out the numbers on the other side (shown in purple below) by realising that you must
<u>divide both sides by 48</u> and then <u>multiply both sides by 60</u>. It's tricky.

÷48 { 48 g of Mg .... 80 g of MgO } ÷48
1 g of Mg .... 1.67 g of MgO
×60 { 60 g of Mg .... 100 g of MgO } ×60

The mass of product is called the <u>yield</u> of a reaction.
<u>In practice</u> you never get 100% of the yield, so the amount of product will be <u>slightly less than calculated</u> (see p. 76).

This finally tells us that <u>60 g of magnesium will produce 100 g of magnesium oxide</u>.
If the question had said "Find how much magnesium gives 500 g of magnesium oxide", you'd fill in the
MgO side first, <u>because that's the one you'd have the information about</u>. Got it? Good-O!

Reaction mass calculations — no worries, matey...

The only way to get good at these is to practise. So have a go at these: Answers on page 140.
1) Find the mass of calcium which gives 30 g of calcium oxide (CaO) when burnt in air.
2) What mass of fluorine fully reacts with potassium to make 116 g of potassium fluoride (KF)?

Atom Economy

It's important in industrial reactions that as much of the reactants as possible get turned into useful products. This depends on the <u>atom economy</u> and the <u>percentage yield</u> (see next page) of the reaction.

"Atom Economy" — % of Reactants Changed to Useful Products

1) A lot of reactions make <u>more than one product</u>. Some of them will be <u>useful</u>, but others will just be <u>waste</u>, e.g. when you make calcium oxide from limestone, you also get CO_2 as a waste product.

2) The <u>atom economy</u> of a reaction tells you how much of the <u>mass</u> of the reactants is wasted when manufacturing a chemical. Here's the equation:

$$\text{atom economy} = \frac{\text{total } M_r \text{ of desired products}}{\text{total } M_r \text{ of all products}} \times 100$$

3) <u>100%</u> atom economy means that <u>all</u> the atoms in the reactants have been turned into <u>useful</u> (desired) <u>products</u>. The <u>higher</u> the atom economy the '<u>greener</u>' the process.

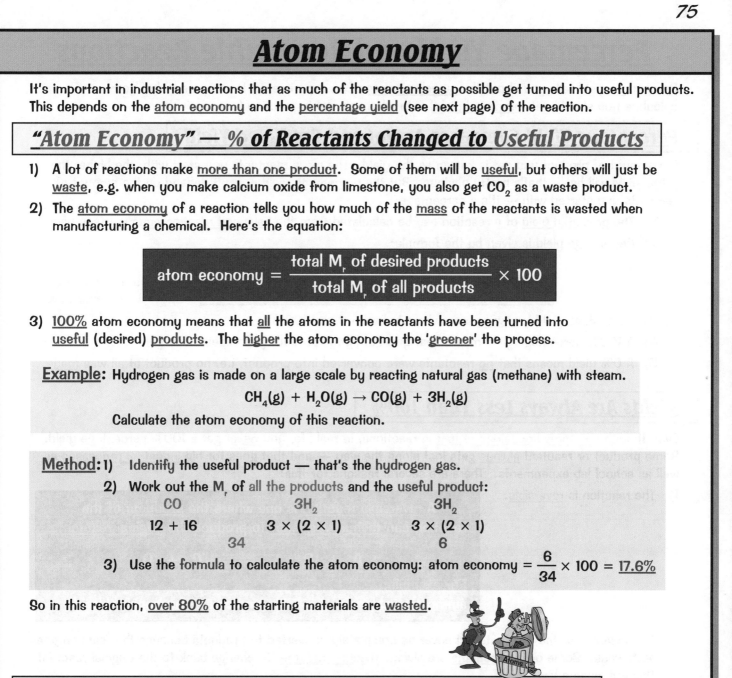

<u>Example:</u> Hydrogen gas is made on a large scale by reacting natural gas (methane) with steam.

$$CH_4(g) + H_2O(g) \rightarrow CO(g) + 3H_2(g)$$

Calculate the atom economy of this reaction.

<u>Method:</u> 1) Identify the useful product — that's the hydrogen gas.

2) Work out the M_r of all the products and the useful product:

CO	$3H_2$	$3H_2$
12 + 16	3 × (2 × 1)	3 × (2 × 1)
34		6

3) Use the formula to calculate the atom economy: $\text{atom economy} = \frac{6}{34} \times 100 = \underline{17.6\%}$

So in this reaction, <u>over 80%</u> of the starting materials are <u>wasted</u>.

High Atom Economy is Better for Profits and the Environment

1) Pretty obviously, if you're making <u>lots of waste</u>, that's a <u>problem</u>.

2) Reactions with low atom economy <u>use up resources</u> very quickly. At the same time, they make lots of <u>waste</u> materials that have to be <u>disposed</u> of somehow. That tends to make these reactions <u>unsustainable</u> — the raw materials will run out and the waste has to go somewhere.

3) For the same reasons, low atom economy reactions aren't usually <u>profitable</u>. Raw materials are <u>expensive to buy</u>, and waste products can be expensive to <u>remove</u> and dispose of <u>responsibly</u>.

4) The best way around the problem is to find a <u>use</u> for the waste products rather than just <u>throwing them away</u>. There's often <u>more than one way</u> to make the product you want, so the trick is to come up with a reaction that gives <u>useful "by-products"</u> rather than useless ones.

5) The reactions with the <u>highest</u> atom economy are the ones that only have <u>one product</u>. Those reactions have an atom economy of <u>100%</u>.

Atom economy — important but not the whole story...

The stuff on this page applies to any industrial reaction. In the real world, high atom economy isn't enough, though. We have to think about the <u>percentage yield</u> of the reaction (next page) and the <u>energy cost</u> as well.

Percentage Yield and Reversible Reactions

Percentage yield tells you about the <u>overall success</u> of an experiment. It compares what you calculate you should get (<u>predicted yield</u>) with what you get in practice (<u>actual yield</u>).

Percentage Yield Compares Actual and Predicted Yield

The amount of product you get is known as the <u>yield</u>. The more reactants you start with, the higher the <u>actual yield</u> will be — that's pretty obvious. But the <u>percentage yield doesn't</u> depend on the amount of reactants you started with — it's a <u>percentage</u>.

1) The <u>predicted yield</u> of a reaction can be calculated from the <u>balanced reaction equation</u> (see page 14).

2) Percentage yield is given by the formula:

$$\text{percentage yield} = \frac{\text{actual yield (grams)}}{\text{predicted yield (grams)}} \times 100$$

(The predicted yield is sometimes called the theoretical yield.)

3) Percentage yield is <u>always</u> somewhere between 0 and 100%.

4) A 100% percentage yield means that you got <u>all</u> the product you expected to get.

5) A 0% yield means that <u>no</u> reactants were converted into product, i.e. no product at all was <u>made</u>.

Yields Are Always Less Than 100%

Even though <u>no atoms are gained or lost</u> in reactions, in real life, you <u>never</u> get a 100% percentage yield. Some product or reactant <u>always</u> gets lost along the way — and that goes for big <u>industrial processes</u> as well as school lab experiments. There are several reasons for this:

1) The reaction is <u>reversible</u>:

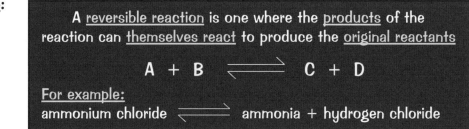

A <u>reversible reaction</u> is one where the <u>products</u> of the reaction can <u>themselves react</u> to produce the <u>original reactants</u>

$$A + B \rightleftharpoons C + D$$

<u>For example:</u>
ammonium chloride $\rightleftharpoons$ ammonia + hydrogen chloride

    This means that the reactants will never be completely converted to products because the reaction goes both ways. Some of the <u>products</u> are always <u>reacting together</u> to change back to the original reactants. This will mean a <u>lower yield</u>.

2) When you <u>filter a liquid</u> to remove <u>solid particles</u>, you nearly always <u>lose</u> a bit of liquid or a bit of solid. So, some of the product may be lost when it's <u>separated</u> from the reaction mixture.

3) You always lose a bit of liquid when you <u>transfer</u> it from one container to another — even if you manage not to spill it. Some of it always gets left behind on the <u>inside surface</u> of the old container.

4) Liquids evaporate <u>all the time</u> — and even more so while they're being heated.

5) Things don't always go exactly to plan. Sometimes there can be other <u>unexpected reactions</u> happening which <u>use up the reactants</u>. This means there's not as much reactant to make the <u>product</u> you want.

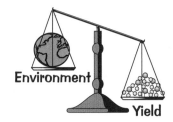

Environment

Yield

Thinking about product yield is important for <u>sustainable development</u>. Sustainable development is about making sure that we don't use <u>resources</u> faster than they can be <u>replaced</u> — there needs to be enough for <u>future generations</u> too. So, for example, using as <u>little energy</u> as possible to create the <u>highest product yield possible</u> means that resources are <u>saved</u>. A low yield means wasted chemicals — not very sustainable.

You can't always get what you want...

A high percentage yield means there's <u>not much waste</u> — good for <u>preserving resources</u> and keeping <u>costs down</u>. If a reaction's worth doing commercially, it generally has to have a high percentage yield or recyclable reactants.

Revision Summary for Section Seven

Don't know if you've noticed but, right on cue, revision summaries have a nasty habit of popping up at the end of every section in this book. I've made a lot of effort to make sure that this happens so it'd be plain rude of you to just skip past them... So mind your manners and have a go at the one below.

1) Define relative formula mass.

2)* Find A_r or M_r for these (use the periodic table at the front of the book):
   a) Ca   b) Ag   c) CO_2 d) $MgCO_3$ e) Na_2CO_3 f) ZnO   g) KOH   h) NH_3

3) What is the link between moles and relative formula mass?

4)* a) Calculate the percentage mass of carbon in:
      i) $CaCO_3$ ii) CO_2 iii) CH_4
   b) Calculate the percentage mass of metal in:
      i) Na_2O ii) Fe_2O_3 iii) Al_2O_3

5) a) What is an empirical formula?
   b)*Find the empirical formula of the compound formed when
      21.9 g of magnesium, 29.2 g of sulfur and 58.4 g of oxygen react.

6) What volume does one mole of gas take up at room temperature and pressure?

7)* What mass of sodium is needed to produce 108.2 g of sodium oxide (Na_2O)?

8) Write the equation for calculating the atom economy of a reaction.

9) Explain why it is important to use industrial reactions with a high atom economy.

10) Describe three factors that can reduce the percentage yield of a reaction.

* Answers on page 140.

Section Seven — Equations and Calculations

Hazards and State Symbols

Safety is an important thing — it seems everything is out to get you in a lab...

Hazard Symbols *Are There To* Warn *You About* Danger

Lots of the chemicals you'll meet in Chemistry can be bad for you or dangerous in some way. That's why the chemical containers will normally have symbols on them to tell you what the dangers are. Understanding these hazard symbols means that you'll be able to use suitable safe-working procedures in the lab.

Oxidising
Provides oxygen which allows other materials to burn more fiercely.
Example: Liquid oxygen.

Harmful
Like toxic but not quite as dangerous.
Example: Copper sulfate.

Highly Flammable
Catches fire easily.
Example: Petrol.

Explosive
Can explode — BANG.
Example: Some peroxides.

Toxic
Can cause death either by swallowing, breathing in, or absorption through the skin.
Example: Hydrogen cyanide.

Corrosive
Attacks and destroys living tissues, including eyes and skin.
Example: Concentrated sulfuric acid.

Irritant
Not corrosive but can cause reddening or blistering of the skin.
Examples: Bleach, children, etc.

State Symbols *Tell You What* Physical State *It's In*

These are easy enough, so make sure you know them — especially aq (aqueous).

| (s) — Solid | (l) — Liquid | (g) — Gas | (aq) — Dissolved in water |

E.g. $2Mg_{(s)} + O_{2(g)} \rightarrow 2MgO_{(s)}$

No, it means 'oxidising' — not a guy with a wacky hairstyle...

The stuff on this page is all pretty important when you're doing experiments with chemicals in the lab. Hazard symbols that appear on the containers of chemicals aren't just there to look pretty and neither are state symbols. Think about it — knowing whether you're dealing with an explosive gas or a corrosive liquid is going to be pretty helpful when it comes to choosing the right equipment.

Acids and Alkalis

Testing the pH of a solution means using an <u>indicator</u> — and that means pretty <u>colours</u>...

The pH Scale Goes From 0 to 14

1) The <u>pH scale</u> is a measure of how <u>acidic</u> or <u>alkaline</u> a solution is.

2) The <u>strongest acid</u> has <u>pH 0</u>. The <u>strongest alkali</u> has <u>pH 14</u>.

3) A <u>neutral</u> substance has <u>pH 7</u> (e.g. pure water).

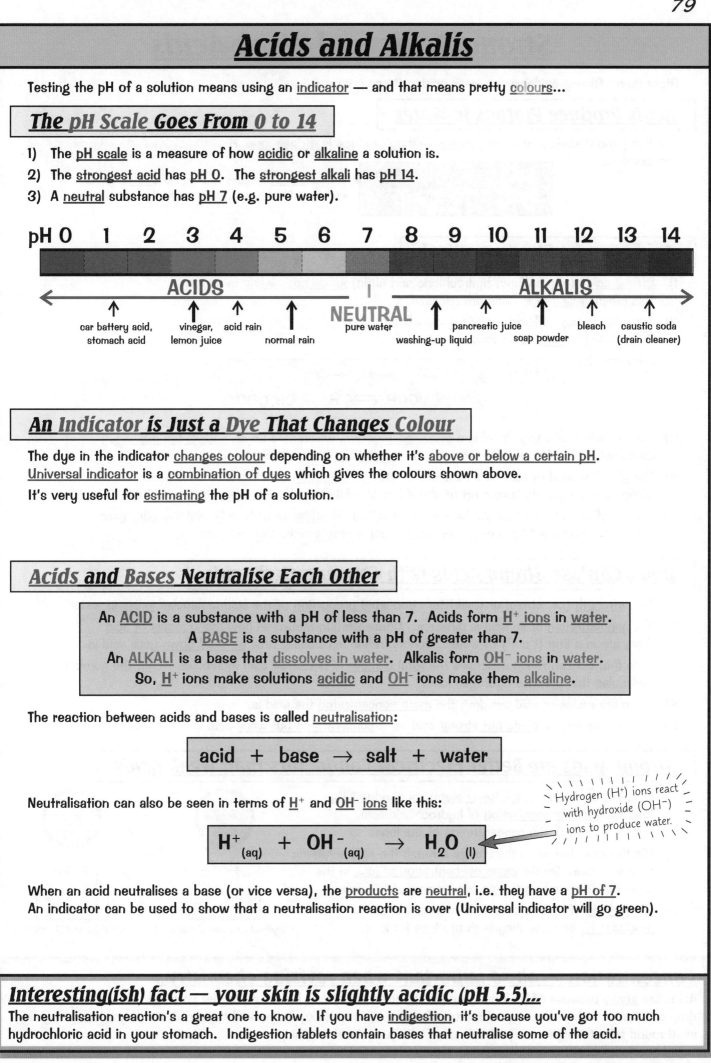

pH 0 1 2 3 4 5 6 7 8 9 10 11 12 13 14

← ACIDS | ALKALIS →

NEUTRAL

car battery acid, stomach acid   vinegar, lemon juice   acid rain   normal rain   pure water   washing-up liquid   pancreatic juice   soap powder   bleach   caustic soda (drain cleaner)

An Indicator is Just a Dye That Changes Colour

The dye in the indicator <u>changes colour</u> depending on whether it's <u>above or below a certain pH</u>.
<u>Universal indicator</u> is a <u>combination of dyes</u> which gives the colours shown above.

It's very useful for <u>estimating</u> the pH of a solution.

Acids and Bases Neutralise Each Other

> An <u>ACID</u> is a substance with a pH of less than 7. Acids form <u>H^+ ions</u> in <u>water</u>.
> A <u>BASE</u> is a substance with a pH of greater than 7.
> An <u>ALKALI</u> is a base that <u>dissolves in water</u>. Alkalis form <u>OH^- ions</u> in <u>water</u>.
> So, <u>H^+</u> ions make solutions <u>acidic</u> and <u>OH^-</u> ions make them <u>alkaline</u>.

The reaction between acids and bases is called <u>neutralisation</u>:

$$ \text{acid} \ + \ \text{base} \ \rightarrow \ \text{salt} \ + \ \text{water} $$

Neutralisation can also be seen in terms of <u>H^+</u> and <u>OH^- ions</u> like this:

$$ H^+_{(aq)} \ + \ OH^-_{(aq)} \ \rightarrow \ H_2O_{(l)} $$

Hydrogen (H^+) ions react with hydroxide (OH^-) ions to produce water.

When an acid neutralises a base (or vice versa), the <u>products</u> are <u>neutral</u>, i.e. they have a <u>pH of 7</u>.
An indicator can be used to show that a neutralisation reaction is over (Universal indicator will go green).

Interesting(ish) fact — your skin is slightly acidic (pH 5.5)...

The neutralisation reaction's a great one to know. If you have <u>indigestion</u>, it's because you've got too much hydrochloric acid in your stomach. Indigestion tablets contain bases that neutralise some of the acid.

Strong Acids and Weak Acids

Right then. Strong acids versus weak acids. Brace yourself.

Acids Produce Protons in Water

The thing about acids is that they ionise — they produce hydrogen ions, H⁺.
For example,

$$HCl \rightarrow H^+ + Cl^-$$
$$HNO_3 \rightarrow H^+ + NO_3^-$$

An H⁺ ion is just a proton.

But HCl doesn't produce hydrogen ions until it meets water — so hydrogen chloride gas isn't an acid.

Acids Can be Strong or Weak

1) Strong acids (e.g. sulfuric, hydrochloric and nitric) ionise completely in water.
   This means loads of H⁺ ions are released.

2) Weak acids (e.g. ethanoic, citric and carbonic) do not fully ionise.
   Only small numbers of H⁺ ions are released.

   For example,

   Strong acid: $HCl \longrightarrow H^+ + Cl^-$
   Weak acid: $CH_3COOH \rightleftharpoons H^+ + CH_3COO^-$

   Use a 'reversible reaction' arrow for a weak acid.

3) The ionisation of a weak acid is a reversible reaction, which sets up an equilibrium mixture.
   Since only a few H⁺ ions are released, the equilibrium lies well to the left.

4) The pH of an acid or alkali is a measure of the concentration of H⁺ ions in the solution.
   Strong acids typically have a pH of about 1 or 2, while the pH of a weak acid might be 4, 5 or 6.

5) The pH of an acid or alkali can be measured with a pH meter or with Universal indicator paper
   (or can be estimated by seeing how fast a sample reacts with, say, magnesium).

Don't Confuse Strong Acids with Concentrated Acids

1) Acid strength (i.e. strong or weak) tells you what proportion of the acid molecules ionise in water.

2) The concentration of an acid is different. Concentration measures how many moles of acid
   there are in a litre (1 dm³) of water. Concentration is basically how watered down your acid is.

3) Note that concentration describes the total number of dissolved acid molecules — not the number of
   molecules that produce hydrogen ions.

4) The more moles of acid per dm³, the more concentrated the acid is.

5) So you can have a dilute but strong acid, or a concentrated but weak acid.

Strong Acids are Better Electrical Conductors than Weak Acids

1) Ethanoic acid has a much lower electrical conductivity
   than the same concentration of hydrochloric acid.
   It's all to do with the concentration of the ions.

2) It's the ions that carry the charge through the acid solutions
   as they move. So the lower concentration of ions in the
   weak acid means less charge can be carried. Simple.

3) Electrolysis of hydrochloric acid or ethanoic acid
   produces H₂ because they both produce H⁺ ions.

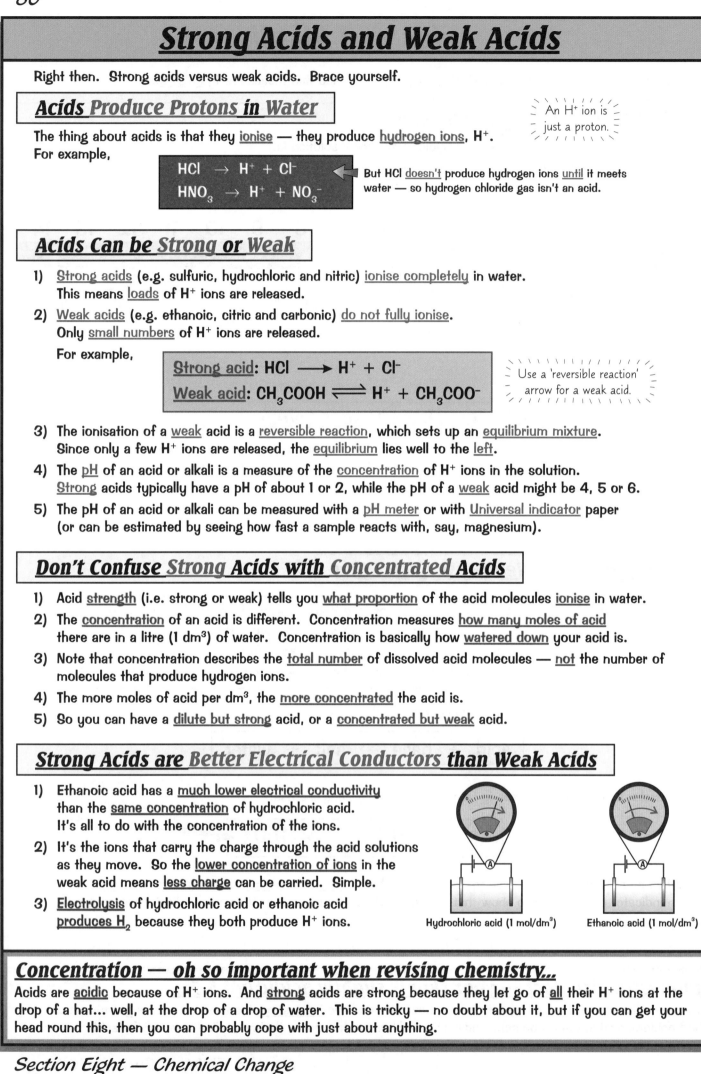

Hydrochloric acid (1 mol/dm³)   Ethanoic acid (1 mol/dm³)

Concentration — oh so important when revising chemistry...

Acids are acidic because of H⁺ ions. And strong acids are strong because they let go of all their H⁺ ions at the
drop of a hat... well, at the drop of a drop of water. This is tricky — no doubt about it, but if you can get your
head round this, then you can probably cope with just about anything.

Acids Reacting With Metals

Sadly, the salts on this page aren't the sort you'd want to go putting on your fish 'n' chips.

Metals React With Acids to Give Salts

Acid + Metal → Salt + Hydrogen

That's written big 'cos it's kinda worth remembering. Here's the typical experiment:

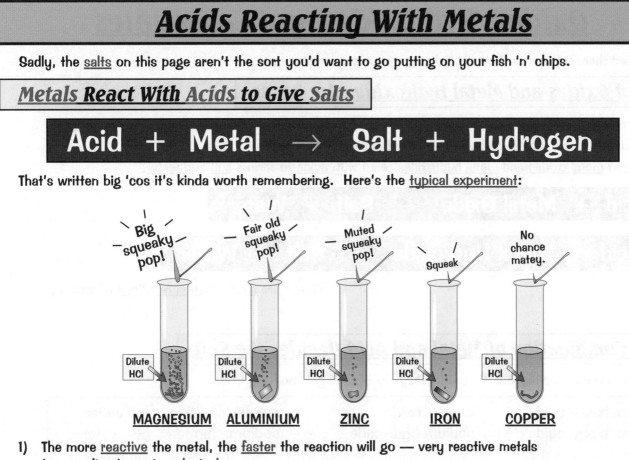

Big squeaky pop!   Fair old squeaky pop!   Muted squeaky pop!   Squeak   No chance matey.

Dilute HCl — MAGNESIUM   ALUMINIUM   ZINC   IRON   COPPER

1) The more reactive the metal, the faster the reaction will go — very reactive metals (e.g. sodium) react explosively.

2) Copper does not react with dilute acids at all — because it's less reactive than hydrogen.

3) The speed of reaction is indicated by the rate at which the bubbles of hydrogen are given off.

4) The hydrogen is confirmed by the burning splint test giving the notorious 'squeaky pop'.

5) The name of the salt produced depends on which metal is used, and which acid is used:

Hydrochloric Acid Will Always Produce Chloride Salts:

$2HCl + Mg \rightarrow MgCl_2 + H_2$ (Magnesium chloride)
$6HCl + 2Al \rightarrow 2AlCl_3 + 3H_2$ (Aluminium chloride)
$2HCl + Zn \rightarrow ZnCl_2 + H_2$ (Zinc chloride)

Sulfuric Acid Will Always Produce Sulfate Salts:

$H_2SO_4 + Mg \rightarrow MgSO_4 + H_2$ (Magnesium sulfate)
$3H_2SO_4 + 2Al \rightarrow Al_2(SO_4)_3 + 3H_2$ (Aluminium sulfate)
$H_2SO_4 + Zn \rightarrow ZnSO_4 + H_2$ (Zinc sulfate)

Nitric Acid Produces Nitrate Salts When NEUTRALISED, But...

Nitric acid reacts fine with alkalis, to produce nitrates, but it can play silly devils with metals and produce nitrogen oxides instead, so we'll ignore it here. Chemistry's a real messy subject sometimes, innit.

Nitric acid, tut — there's always one...

Okay, so this stuff isn't exactly a laugh a minute, but at least it's fairly straightforward learning. Metals that are less reactive than hydrogen don't react with acid, and some metals like sodium and potassium are too reactive to mix with acid in a school lab — your beaker would explode.

Oxides, Hydroxides and Carbonates

I'm afraid there's more stuff on <u>neutralisation</u> reactions coming up...

Metal Oxides and Metal Hydroxides Are Bases

1) Some <u>metal oxides</u> and <u>metal hydroxides</u> dissolve in <u>water</u>. These soluble compounds are <u>alkalis</u>.
2) Even bases that won't dissolve in water will still react with acids.
3) So, all <u>metal oxides</u> and <u>metal hydroxides</u> react with <u>acids</u> to form a <u>salt</u> and <u>water</u>.

> **Acid + Metal Oxide → Salt + Water**

> **Acid + Metal Hydroxide → Salt + Water**

(These are <u>neutralisation reactions</u> of course)

The Combination of Metal and Acid Decides the Salt

This isn't exactly exciting but it's pretty easy, so try and get the hang of it:

hydro**chloric** acid	+	**copper** oxide	→	copper chloride + water
hydro**chloric** acid	+	**sodium** hydroxide	→	sodium chloride + water
sulfuric acid	+	**zinc** oxide	→	zinc sulfate + water
sulfuric acid	+	**calcium** hydroxide	→	calcium sulfate + water
nitric acid	+	**magnesium** oxide	→	magnesium nitrate + water
nitric acid	+	**potassium** hydroxide	→	potassium nitrate + water

The symbol equations are all pretty much the same. Here are two of them:

$$H_2SO_{4\,(aq)} + ZnO_{(s)} \rightarrow ZnSO_{4\,(aq)} + H_2O_{(l)}$$
$$HNO_{3\,(aq)} + KOH_{(aq)} \rightarrow KNO_{3\,(aq)} + H_2O_{(l)}$$

Metal Carbonates Give Salt + Water + Carbon Dioxide

More gripping reactions involving acids. At least there are some <u>bubbles</u> involved here.

> **Acid + Metal Carbonate → Salt + Water + Carbon Dioxide**

The reaction is the same as any other neutralisation reaction EXCEPT that <u>carbonates</u> give off <u>carbon dioxide</u> as well. Here's an example for you:

hydro**chloric** acid + sodium carbonate → sodium chloride + **water** + carbon dioxide
$$2HCl_{(aq)} + Na_2CO_{3(s)} \rightarrow 2NaCl_{(aq)} + H_2O_{(l)} + CO_{2(g)}$$

Someone threw some NaCl at me — I said, "Hey that's a salt"...

The acid + carbonate reaction is one you might have to do at home. If you live in a <u>hard water</u> area, you'll get insoluble $MgCO_3$ and $CaCO_3$ 'furring up' your kettle. You can get rid of this with 'descaler', which is dilute <u>acid</u> (often citric acid) — this reacts with the <u>insoluble carbonates</u> to make <u>soluble salts</u>.

Making Insoluble Salts

Some salts are <u>soluble</u> and some are <u>insoluble</u> — it's just the way the cookie crumbles.

The Rules of Solubility

This table is a pretty fail-safe way of working out whether a substance is soluble in water or not.

Substance	Soluble or Insoluble?
common salts of sodium, potassium and ammonium	soluble
nitrates	soluble
common chlorides	soluble (except silver chloride and lead chloride)
common sulfates	soluble (except lead, barium and calcium sulfate)
common carbonates and hydroxides	insoluble (except for sodium, potassium and ammonium ones)

Making Insoluble Salts — Precipitation Reactions

1) To make a pure, dry sample of an <u>insoluble</u> salt, you can use a <u>precipitation reaction</u>.
   You just need to pick the right two <u>soluble salts</u>, they <u>react</u> and you get your <u>insoluble salt</u>.

2) E.g. to make <u>lead chloride</u> (insoluble), mix <u>lead nitrate</u> and <u>sodium chloride</u> (both soluble).

lead nitrate + sodium chloride → lead chloride + sodium nitrate

$$Pb(NO_3)_{2(aq)} + 2NaCl_{(aq)} \longrightarrow PbCl_{2(s)} + 2NaNO_{3(aq)}$$

Method

Stage 1

1) Add 1 spatula of <u>lead nitrate</u> to a test tube, and fill it with <u>distilled water</u>. <u>Shake it thoroughly</u> to ensure that all the lead nitrate has <u>dissolved</u>. Then do the same with 1 spatula of <u>sodium chloride</u>. (Use distilled water to make sure there are <u>no other ions</u> about.)

2) Tip the <u>two solutions</u> into a small beaker, and give it a good stir to make sure it's all mixed together. The lead chloride should <u>precipitate</u> out.

precipitate

Stage 2

filter paper

filter funnel

1) Put a folded piece of <u>filter paper</u> into a <u>filter funnel</u>, and stick the funnel into a <u>conical flask</u>.

2) <u>Pour</u> the contents of the beaker into the middle of the filter paper. (Make sure that the solution doesn't go above the filter paper — otherwise some of the solid could dribble down the side.)

3) <u>Swill out</u> the beaker with more distilled water, and tip this into the filter paper — to make sure you get <u>all the product</u> from the beaker.

Stage 3

1) Rinse the contents of the filter paper with distilled water to make sure that <u>all the soluble sodium nitrate</u> has been washed away.

2) Then just scrape the <u>lead chloride</u> onto fresh filter paper and leave to dry.

lead chloride

If you aren't part of the solution, you're part of the precipitate...

You can use the <u>solubility rules</u> to predict whether a <u>precipitate</u> will be formed when two given solutions are mixed together — and to work out its <u>name</u>. It's almost as impressive as a crystal ball — only not quite.

Making Soluble Salts

When an acid and an alkali react you get a <u>salt</u>. However, if the salt that's made is <u>soluble</u>, getting hold of it can be tricky. That's where a bit of cunning comes in...

Making <u>Soluble Salts</u> Using an <u>Acid</u> and an <u>Insoluble Reactant</u>

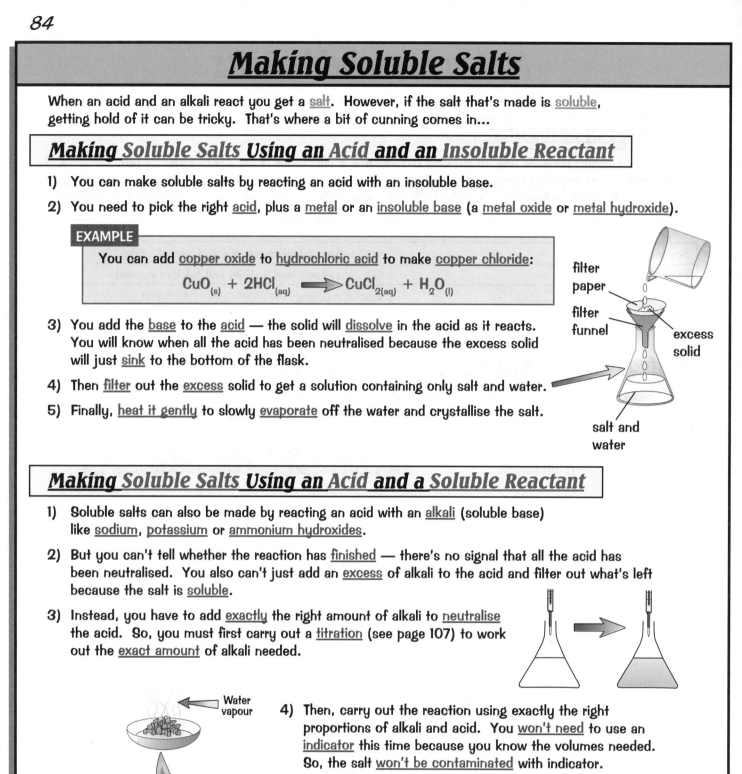

1) You can make soluble salts by reacting an acid with an insoluble base.

2) You need to pick the right <u>acid</u>, plus a <u>metal</u> or an <u>insoluble base</u> (a <u>metal oxide</u> or <u>metal hydroxide</u>).

> **EXAMPLE**
> You can add <u>copper oxide</u> to <u>hydrochloric acid</u> to make <u>copper chloride</u>:
> $$CuO_{(s)} + 2HCl_{(aq)} \longrightarrow CuCl_{2(aq)} + H_2O_{(l)}$$

filter paper

filter funnel

excess solid

3) You add the <u>base</u> to the <u>acid</u> — the solid will <u>dissolve</u> in the acid as it reacts. You will know when all the acid has been neutralised because the excess solid will just <u>sink</u> to the bottom of the flask.

4) Then <u>filter</u> out the <u>excess</u> solid to get a solution containing only salt and water.

5) Finally, <u>heat it gently</u> to slowly <u>evaporate</u> off the water and crystallise the salt.

salt and water

Making <u>Soluble Salts</u> Using an <u>Acid</u> and a <u>Soluble Reactant</u>

1) Soluble salts can also be made by reacting an acid with an <u>alkali</u> (soluble base) like <u>sodium</u>, <u>potassium</u> or <u>ammonium hydroxides</u>.

2) But you can't tell whether the reaction has <u>finished</u> — there's no signal that all the acid has been neutralised. You also can't just add an <u>excess</u> of alkali to the acid and filter out what's left because the salt is <u>soluble</u>.

3) Instead, you have to add <u>exactly</u> the right amount of alkali to <u>neutralise</u> the acid. So, you must first carry out a <u>titration</u> (see page 107) to work out the <u>exact amount</u> of alkali needed.

Water vapour

4) Then, carry out the reaction using exactly the right proportions of alkali and acid. You <u>won't need</u> to use an <u>indicator</u> this time because you know the volumes needed. So, the salt <u>won't be contaminated</u> with indicator.

5) The <u>solution</u> that remains when the reaction is complete contains only the <u>salt</u> and <u>water</u>. <u>Evaporate</u> off the water slowly and you'll be left with a <u>pure</u>, <u>dry</u> salt.

> **EXAMPLE**
> <u>Sulfuric acid</u> can be reacted with <u>sodium hydroxide</u> to make <u>sodium sulfate</u>.
> $$H_2SO_{4(aq)} + 2NaOH_{(aq)} \longrightarrow Na_2SO_{4(aq)} + 2H_2O_{(l)}$$

<u>Get two beakers, mix 'em together — job's a good'un...</u>

If you're making a soluble salt, you need to think <u>carefully</u> about what <u>chemicals</u> you'd need to get the salt you want and what <u>method</u> you'd use. If you're making the salt from an acid and a <u>soluble</u> reactant then you first need to carry out a <u>titration</u> so you know exactly how much of the reactant to use. If you're making the salt using an <u>insoluble</u> reactant then you add a load to the acid and <u>filter</u> out any that's left at the end of the reaction. In both cases you're left with a solution containing only <u>salt</u> and <u>water</u>.

Redox Reactions

In chemistry, things get oxidised and reduced all the time.

If Electrons are Transferred, It's a Redox Reaction

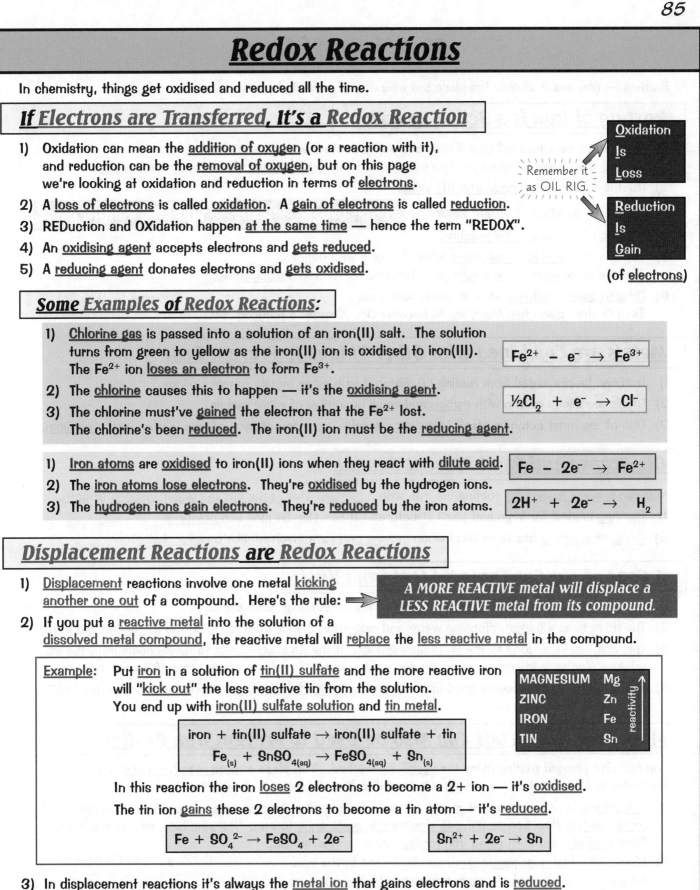

1) Oxidation can mean the <u>addition of oxygen</u> (or a reaction with it), and reduction can be the <u>removal of oxygen</u>, but on this page we're looking at oxidation and reduction in terms of <u>electrons</u>.

2) A <u>loss of electrons</u> is called <u>oxidation</u>. A <u>gain of electrons</u> is called <u>reduction</u>.

3) REDuction and OXidation happen <u>at the same time</u> — hence the term "REDOX".

4) An <u>oxidising agent</u> accepts electrons and <u>gets reduced</u>.

5) A <u>reducing agent</u> donates electrons and <u>gets oxidised</u>.

Remember it as OIL RIG.

Oxidation **I**s **L**oss

Reduction **I**s **G**ain

(of <u>electrons</u>)

Some Examples of Redox Reactions:

1) <u>Chlorine gas</u> is passed into a solution of an iron(II) salt. The solution turns from green to yellow as the iron(II) ion is oxidised to iron(III). The Fe^{2+} ion <u>loses an electron</u> to form Fe^{3+}.

$$Fe^{2+} - e^- \rightarrow Fe^{3+}$$

2) The <u>chlorine</u> causes this to happen — it's the <u>oxidising agent</u>.

3) The chlorine must've <u>gained</u> the electron that the Fe^{2+} lost. The chlorine's been <u>reduced</u>. The iron(II) ion must be the <u>reducing agent</u>.

$$\tfrac{1}{2}Cl_2 + e^- \rightarrow Cl^-$$

1) <u>Iron atoms</u> are <u>oxidised</u> to iron(II) ions when they react with <u>dilute acid</u>.

$$Fe - 2e^- \rightarrow Fe^{2+}$$

2) The <u>iron atoms lose electrons</u>. They're <u>oxidised</u> by the hydrogen ions.

3) The <u>hydrogen ions gain electrons</u>. They're <u>reduced</u> by the iron atoms.

$$2H^+ + 2e^- \rightarrow H_2$$

Displacement Reactions are Redox Reactions

1) <u>Displacement</u> reactions involve one metal <u>kicking another one out</u> of a compound. Here's the rule:

A MORE REACTIVE metal will displace a LESS REACTIVE metal from its compound.

2) If you put a <u>reactive metal</u> into the solution of a <u>dissolved metal compound</u>, the reactive metal will <u>replace</u> the <u>less reactive metal</u> in the compound.

Example: Put <u>iron</u> in a solution of <u>tin(II) sulfate</u> and the more reactive iron will "<u>kick out</u>" the less reactive tin from the solution. You end up with <u>iron(II) sulfate solution</u> and <u>tin metal</u>.

MAGNESIUM	Mg	
ZINC	Zn	reactivity
IRON	Fe	
TIN	Sn	

iron + tin(II) sulfate → iron(II) sulfate + tin

$$Fe_{(s)} + SnSO_{4(aq)} \rightarrow FeSO_{4(aq)} + Sn_{(s)}$$

In this reaction the iron <u>loses</u> 2 electrons to become a 2+ ion — it's <u>oxidised</u>.

The tin ion <u>gains</u> these 2 electrons to become a tin atom — it's <u>reduced</u>.

$$Fe + SO_4^{2-} \rightarrow FeSO_4 + 2e^-$$

$$Sn^{2+} + 2e^- \rightarrow Sn$$

3) In displacement reactions it's always the <u>metal ion</u> that gains electrons and is <u>reduced</u>. The <u>metal atom</u> always loses electrons and is <u>oxidised</u>.

4) In the exam you could be asked to write <u>word or symbol equations</u> to show displacement reactions. If you're ever asked to <u>predict</u> whether or not a displacement reaction will happen, all you have to remember is that <u>more reactive metals displace less reactive ones</u> and you'll be fine and dandy.

REDOX — great for bubble baths. Oh no, wait...

Try writing some displacement reaction equations now — write the equation for the reaction between <u>zinc and iron chloride</u> ($FeCl_2$). What's being <u>oxidised</u>? What's being <u>reduced</u>? Practise till you can <u>do it in your sleep</u>.

86

Rusting of Iron

Rusting — you see it all over the place but why does it actually happen...

Rusting of Iron is a Redox Reaction

1) Iron and some steels will <u>rust</u> if they come into contact with air and water.
   Rusting only happens when the iron's in contact with <u>both oxygen</u> (from the air) and <u>water</u>.

2) Rust is a form of <u>hydrated iron(III) oxide</u>.

3) Here's the <u>equation for rust</u>: ➡ iron + oxygen + water → hydrated iron(III) oxide

4) Rusting of iron is a <u>redox reaction</u>.

5) This is why. <u>Iron loses electrons</u> when it reacts with oxygen.
   Each Fe atom <u>loses three electrons</u> to become Fe^{3+}. Iron's <u>oxidised</u>.

6) <u>Oxygen gains electrons</u> when it reacts with iron.
   Each O atom <u>gains two electrons</u> to become O^{2-}. Oxygen's <u>reduced</u>.

Remember <u>OIL RIG</u>.

Metals are Combined with Other Things to Prevent Rust

1) Iron can be prevented from rusting by mixing it with <u>other metals</u> to make alloys.

2) <u>Steels</u> are alloys of iron with <u>carbon</u> and small quantities of other metals.

3) One of the most common steels is <u>stainless steel</u> — a rustproof alloy of iron, carbon and <u>chromium</u>.

Oil, Grease and Paint Prevent Rusting

You can <u>prevent rusting</u> by coating the iron with a <u>barrier</u>. This <u>keeps out the water</u>, <u>oxygen</u> or <u>both</u>.

1) <u>Painting</u> is ideal for large and small structures. It can also be nice and <u>colourful</u>.

2) <u>Oiling</u> or <u>greasing</u> has to be used when <u>moving parts</u> are involved, like on <u>bike chains</u>.

A Coat of Tin Can Protect Steel from Rust

1) <u>Tin plating</u> is where a coat of tin is applied to the object, e.g. food cans.

2) The tin acts as a <u>barrier</u>, stopping water and oxygen in the air from reaching the <u>surface</u> of the iron.

3) This only works as long as the <u>tin remains intact</u>. If the tin is <u>scratched</u> to reveal some iron, the <u>iron will lose electrons</u> in <u>preference</u> to the tin and the iron will rust even faster than if it was on its own.

4) That's why it's <u>not</u> always a good idea to buy the <u>reduced bashed tins</u> of food at the supermarket. They could be starting to <u>rust</u>.

More Reactive Metals Can Also be Used to Prevent Iron Rusting

You can also prevent rusting using the <u>sacrificial</u> method. You place a <u>more reactive metal</u> with the iron. The water and oxygen then react with this "sacrificial" metal instead of with the iron.

1) <u>Galvanising</u> is where a coat of <u>zinc</u> is put onto the object. The zinc acts as sacrificial protection — it's <u>more reactive</u> than iron so it'll <u>lose electrons in preference</u> to iron. The zinc also acts as a barrier. Steel <u>buckets</u> and <u>corrugated iron roofing</u> are often galvanised.

2) Blocks of metal, e.g. <u>magnesium</u>, can be bolted to the iron. Magnesium will <u>lose electrons in preference to iron</u>. It's used on the hulls of <u>ships</u>, or on <u>underground iron pipes</u>.

Galvanising protects the metal underneath even when the zinc gets scratched.

<u>Don't get confused</u> about sacrificial protection — it's <u>not a displacement reaction</u>. There isn't a metal reacting with a metal salt — oxygen's reacting with a more reactive metal instead of a less reactive one.

Alloy there Jim Lad...

<u>Rust</u> is one of those really annoying things. It eats your bike, your car, your ship... but then doesn't touch that lovely woolly cardigan that your gran gave you. On the plus side though, you can use it to dye your clothes, just place a rusty object on the fabric, add a splash of vinegar, and voilà — a beautiful orange stain. Marvellous.

Section Eight — Chemical Change

Revision Summary for Section Eight

Have a go at these questions and see how much you can remember. If you're not sure about any of them, don't just skulk past them, check back to the relevant page, get it in your head and tackle the question again. It's the best way, I promise. Especially when you're trying to learn a section full of equations and reactions like that one. You <u>think</u> you're reading about neutralisation, but your brain is actually just saying: 'Blah. Blah blah blah, blah blah. Blah.'

1) Give the meaning of this symbol: ☠

2) Write down the state symbol that means 'dissolved in water'.

3) What does the pH scale show?

4) What type of ions are always present in a) acids and   b) alkalis?

5) What is neutralisation? Write down the general equation for neutralisation in terms of ions.

6) What is the difference between the strength of an acid and its concentration?

7) What is the general equation for reacting an acid with a metal?

8) Name a metal that doesn't react at all with dilute acids.

9) What type of salts do hydrochloric acid and sulfuric acid produce?

10) What type of reaction is "acid + metal oxide", or "acid + metal hydroxide"?

11) Suggest a suitable acid and a suitable metal oxide/hydroxide to mix to form the following salts.
a) copper chloride   b) calcium nitrate   c) zinc sulfate
d) magnesium nitrate   e) sodium sulfate   f) potassium chloride

12) Iron chloride can made by mixing iron hydroxide (an insoluble base) with hydrochloric acid. Describe the method you would use to produce pure, solid iron chloride in the lab.

13) How can you tell when a neutralisation reaction is complete if both the base and the salt are soluble in water?

14) Fill in the gaps: A loss of electrons is _____. A gain of electrons is _____.

15) Give a symbol half-equation for the oxidation of Fe^{2+} to Fe^{3+}.

16) What is a displacement reaction?

17) Give the word equation for the rusting of iron.

18) Explain how greasing and painting protect against rust.

19) Why isn't it always a good plan to buy dented cans of beans?

20) An oil drilling platform uses sacrificial protection. What's "sacrificial protection"?

Rate of Reaction

Reactions can be <u>fast</u> or <u>slow</u> — you've probably already realised that. This page is about what affects the <u>rate of a reaction</u>, and the next page tells you what you can do to <u>measure it</u>. It's exciting stuff. Honest.

Reactions Can Go at All Sorts of *Different Rates*

1) One of the <u>slowest</u> is the <u>rusting</u> of iron (it's not slow enough though — what about my little MGB).

2) A <u>moderate speed</u> reaction is a <u>metal</u> (like magnesium) reacting with <u>acid</u> to produce a gentle stream of <u>bubbles</u>.

3) A <u>really fast</u> reaction is an <u>explosion</u>, where it's all over in a <u>fraction</u> of a second.

The *Rate of a Reaction* Depends on *Four Things:*

1) <u>Temperature</u>
2) <u>Concentration</u>   — (or <u>pressure</u> for gases)
3) <u>Catalyst</u>
4) <u>Surface area of solids</u>   — (or <u>size</u> of solid pieces)

Typical Graphs *for Rate of Reaction*

The plot below shows how the rate of a particular reaction varies under <u>different conditions</u>. The <u>quickest reaction</u> is shown by the line with the <u>steepest slope</u>. Also, the faster a reaction goes, the sooner it finishes, which means that the line becomes <u>flat</u> earlier.

1) <u>Graph 1</u> represents the original <u>fairly slow</u> reaction. The graph is not too steep.

2) <u>Graphs 2 and 3</u> represent the reaction taking place <u>quicker</u> but with the <u>same initial amounts</u>. The slope of the graphs gets steeper.

3) The <u>increased rate</u> could be due to <u>any</u> of these:

> a) increase in <u>temperature</u>
> b) increase in <u>concentration</u> (or pressure)
> c) <u>catalyst</u> added
> d) solid reactant crushed up into <u>smaller bits</u>.

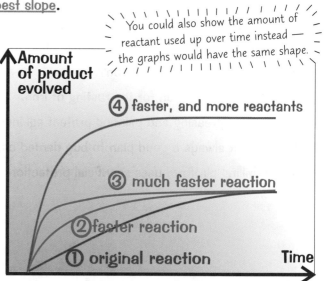

You could also show the amount of reactant used up over time instead — the graphs would have the same shape.

Amount of product evolved

④ faster, and more reactants

③ much faster reaction

② faster reaction

① original reaction

Time

4) <u>Graph 4</u> produces <u>more product</u> as well as going <u>faster</u>. This can <u>only</u> happen if <u>more reactant(s)</u> are added at the start. <u>Graphs 1, 2 and 3</u> all converge at the same level, showing that they all produce the same amount of product, although they take <u>different</u> times to get there.

How to get a fast, furious reaction — crack a wee joke...

<u>Industrial</u> reactions generally use a <u>catalyst</u> and are done at <u>high temperature and pressure</u>. Time is money, so the faster an industrial reaction goes the better... but only <u>up to a point</u>. Chemical plants are quite expensive to rebuild if they get blown into lots and lots of teeny tiny pieces.

Measuring Rates of Reaction

Ways to Measure the Rate of a Reaction

The rate of a reaction can be observed either by measuring how quickly the reactants are used up or how quickly the products are formed. It's usually a lot easier to measure products forming.

The rate of reaction can be calculated using the following formula:

$$\text{Rate of Reaction} = \frac{\text{Amount of reactant used or amount of product formed}}{\text{Time}}$$

There are different ways that the rate of a reaction can be measured. Here are three:

1) Precipitation

1) This is when the product of the reaction is a precipitate which clouds the solution.

2) Observe a mark through the solution and measure how long it takes for it to disappear.

3) The quicker the mark disappears, the quicker the reaction.

4) This only works for reactions where the initial solution is rather see-through.

5) The result is very subjective — different people might not agree over the exact point when the mark 'disappears'.

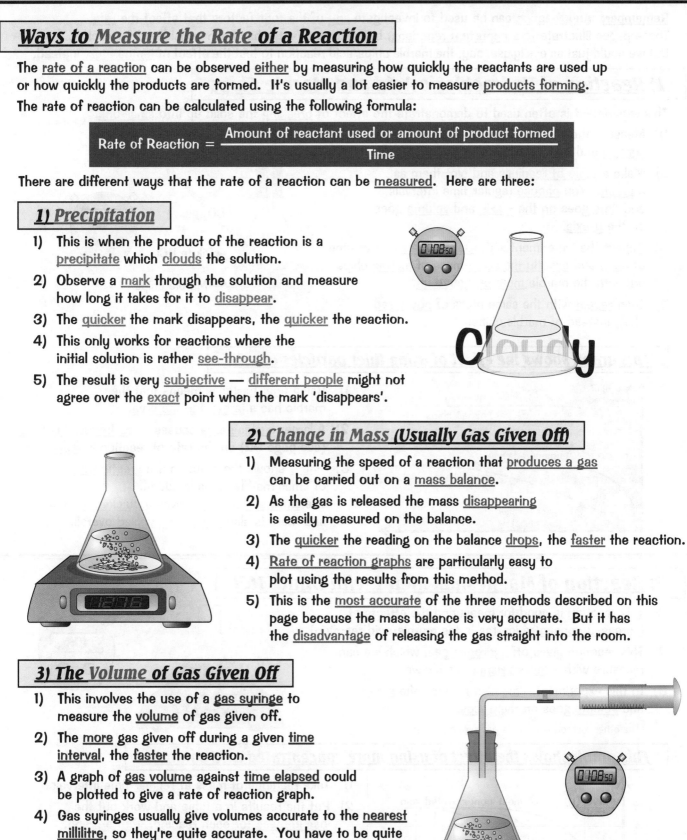

2) Change in Mass (Usually Gas Given Off)

1) Measuring the speed of a reaction that produces a gas can be carried out on a mass balance.

2) As the gas is released the mass disappearing is easily measured on the balance.

3) The quicker the reading on the balance drops, the faster the reaction.

4) Rate of reaction graphs are particularly easy to plot using the results from this method.

5) This is the most accurate of the three methods described on this page because the mass balance is very accurate. But it has the disadvantage of releasing the gas straight into the room.

3) The Volume of Gas Given Off

1) This involves the use of a gas syringe to measure the volume of gas given off.

2) The more gas given off during a given time interval, the faster the reaction.

3) A graph of gas volume against time elapsed could be plotted to give a rate of reaction graph.

4) Gas syringes usually give volumes accurate to the nearest millilitre, so they're quite accurate. You have to be quite careful though — if the reaction is too vigorous, you can easily blow the plunger out of the end of the syringe!

OK have you got your stopwatch ready *BANG!* — oh...

Each method has its pros and cons. The mass balance method is only accurate as long as the flask isn't too hot, otherwise you lose mass by evaporation as well as by the reaction. The first method isn't very accurate, but if you're not producing a gas you can't use either of the other two. Ah well.

Rate of Reaction Experiments

Remember: Any reaction can be used to investigate any of the four factors that affect the rate.
These pages illustrate four important reactions, but only one factor has been considered for each.
But we could just as easily use, say, the marble chips/acid reaction to test the effect of temperature instead.

1) Reaction of Hydrochloric Acid and Marble Chips

This experiment is often used to demonstrate the effect of breaking the solid up into small bits.

1) Measure the volume of gas evolved with a gas syringe and take readings at regular intervals.

2) Make a table of readings and plot them as a graph. You choose regular time intervals, and time goes on the x-axis and volume goes on the y-axis.

3) Repeat the experiment with exactly the same volume of acid, and exactly the same mass of marble chips, but with the marble more crunched up.

4) Then repeat with the same mass of powdered chalk instead of marble chips.

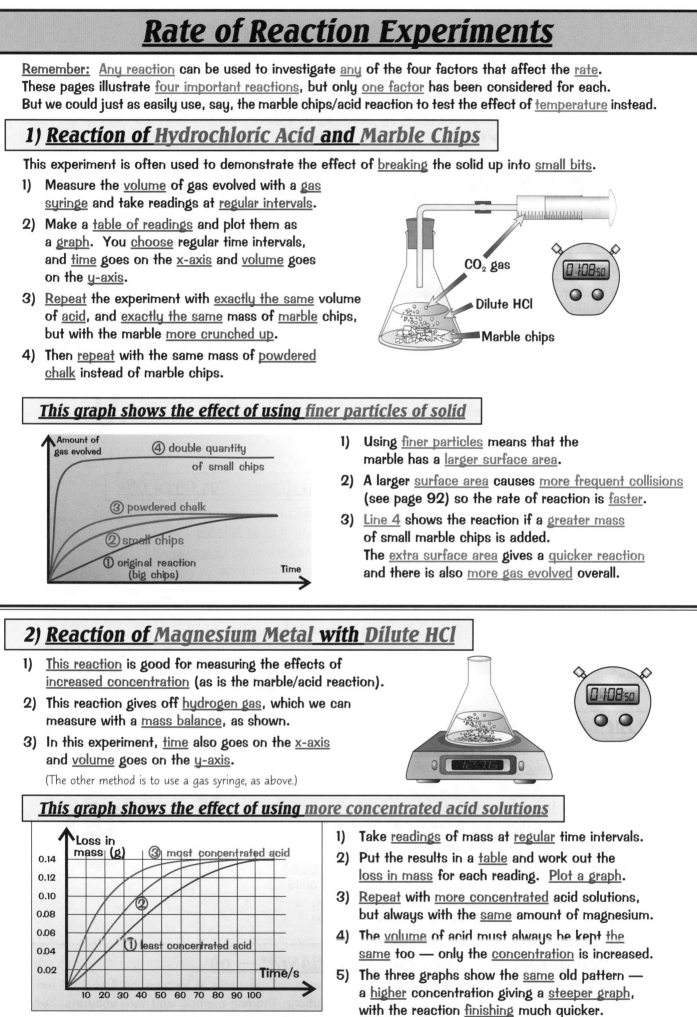

CO₂ gas

Dilute HCl

Marble chips

This graph shows the effect of using finer particles of solid

1) Using finer particles means that the marble has a larger surface area.

2) A larger surface area causes more frequent collisions (see page 92) so the rate of reaction is faster.

3) Line 4 shows the reaction if a greater mass of small marble chips is added. The extra surface area gives a quicker reaction and there is also more gas evolved overall.

2) Reaction of Magnesium Metal with Dilute HCl

1) This reaction is good for measuring the effects of increased concentration (as is the marble/acid reaction).

2) This reaction gives off hydrogen gas, which we can measure with a mass balance, as shown.

3) In this experiment, time also goes on the x-axis and volume goes on the y-axis.

(The other method is to use a gas syringe, as above.)

This graph shows the effect of using more concentrated acid solutions

1) Take readings of mass at regular time intervals.

2) Put the results in a table and work out the loss in mass for each reading. Plot a graph.

3) Repeat with more concentrated acid solutions, but always with the same amount of magnesium.

4) The volume of acid must always be kept the same too — only the concentration is increased.

5) The three graphs show the same old pattern — a higher concentration giving a steeper graph, with the reaction finishing much quicker.

More Rate of Reaction Experiments

3) Sodium Thiosulfate and HCl Produce a Cloudy Precipitate

1) These two chemicals are both <u>clear solutions</u>.

2) They react together to form a <u>yellow precipitate</u> of <u>sulfur</u>.

3) The experiment involves watching a black mark <u>disappear</u> through the <u>cloudy sulfur</u> and <u>timing</u> how long it takes to go.

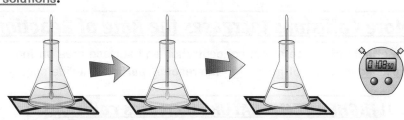

4) The reaction can be <u>repeated</u> for solutions at different <u>temperatures</u>. In practice, that's quite hard to do accurately and safely (it's not a good idea to heat an acid directly). The best way to do it is to use a <u>water bath</u> to heat both solutions to the right temperature <u>before you mix them</u>.

5) The <u>depth</u> of liquid must be kept the <u>same</u> each time, of course.

6) The results will of course show that the <u>higher</u> the temperature the <u>quicker</u> the reaction and therefore the <u>less time</u> it takes for the mark to <u>disappear</u>. These are typical results:

Temperature (°C)	20	25	30	35	40
Time taken for mark to disappear (s)	193	151	112	87	52

This reaction can <u>also</u> be used to test the effects of <u>concentration</u>. One sad thing about this reaction is it <u>doesn't</u> give a set of graphs. Well I think it's sad. All you get is a set of <u>readings</u> of how long it took till the mark disappeared for each temperature. Boring.

4) The Decomposition of Hydrogen Peroxide

This is a <u>good</u> reaction for showing the effect of different <u>catalysts</u>. The decomposition of hydrogen peroxide is:

$$2H_2O_{2\,(aq)} \rightarrow 2H_2O_{(l)} + O_{2\,(g)}$$

1) This is normally quite <u>slow</u> but a sprinkle of <u>manganese(IV) oxide catalyst</u> speeds it up no end. Other catalysts which work are found in: a) <u>potato peel</u> and b) <u>blood</u>.

2) <u>Oxygen gas</u> is given off, which provides an <u>ideal way</u> to measure the rate of reaction using the good ol' <u>gas syringe</u> method.

O₂ gas · Hydrogen peroxide · Catalyst

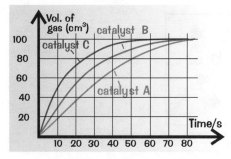

3) Same old graphs of course.

4) <u>Better</u> catalysts give a <u>quicker reaction</u>, which is shown by a <u>steeper graph</u> which levels off quickly.

5) This reaction can also be used to measure the effects of <u>temperature</u>, or of <u>concentration</u> of the H₂O₂ solution. The graphs will look just the same.

BLOOD is a catalyst? — eeurgh...

This stuff's all about comparing those pretty rate of reaction graphs. They tell you the amount of <u>product</u> made (or reactant used up) and the <u>rate of reaction</u>. The <u>steeper</u> the curve, the <u>faster</u> the reaction.

Section Nine — Reaction Rates and Energy Changes

Collision Theory

Reaction rates are explained by collision theory. It's really simple. It just says that the rate of a reaction simply depends on how often and how hard the reacting particles collide with each other. The basic idea is that particles have to collide in order to react, and they have to collide hard enough (with enough energy).

More Collisions Increases the Rate of Reaction

The effects of temperature, concentration and surface area on the rate of reaction can be explained in terms of how often the reacting particles collide successfully.

1) HIGHER TEMPERATURE increases collisions

When the temperature is increased the particles all move quicker.
If they're moving quicker, they're going to collide more often.

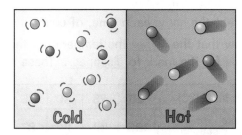

Cold   Hot

2) HIGHER CONCENTRATION (or PRESSURE) increases collisions

If a solution is made more concentrated it means there are more particles of reactant knocking about between the water molecules which makes collisions between the important particles more likely.

In a gas, increasing the pressure means the particles are more squashed up together so there will be more frequent collisions.

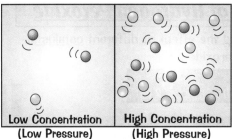

Low Concentration   High Concentration
(Low Pressure)   (High Pressure)

3) LARGER SURFACE AREA increases collisions

If one of the reactants is a solid then breaking it up into smaller pieces will increase the total surface area. This means the particles around it in the solution will have more area to work on, so there'll be more frequent collisions.

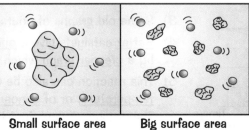

Small surface area   Big surface area

Collision theory — the lamppost ran into me...

Isn't it nice when everything starts to fall into place... The concept's fairly simple — the more often particles bump into each other, and the harder they hit when they do, the faster the reaction happens.

Collision Theory and Catalysts

Without enough activation energy, it's game over before you start.

Faster Collisions Increase the Rate of Reaction

Higher temperature also increases the energy of the collisions, because it makes all the particles move faster.

Increasing the temperature causes faster collisions

Reactions only happen if the particles collide with enough energy.

The minimum amount of energy needed by the particles to react is known as the activation energy.

At a higher temperature there will be more particles colliding with enough energy to make the reaction happen.

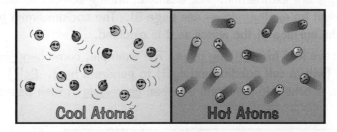

Cool Atoms   Hot Atoms

A Catalyst Increases the Number of Successful Collisions

A catalyst is a substance which increases the speed of a reaction, without being chemically changed or used up in the reaction — and because it isn't used up, you only need a tiny bit of it to catalyse large amounts of reactants. Catalysts tend to be very fussy about which reactions they catalyse though — you can't just stick any old catalyst in a reaction and expect it to work.

A catalyst works by giving the reacting particles a surface to stick to where they can bump into each other — and reduces the energy needed by the particles before they react. So the overall number of collisions isn't increased, but the number of successful collisions is.

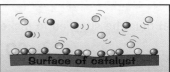

Surface of catalyst

Catalysts Help Reduce Costs in Industrial Reactions

1) Catalysts are very important for commercial reasons — most industrial reactions use them.

2) Catalysts increase the rate of the reaction, which saves a lot of money simply because the plant doesn't need to operate for as long to produce the same amount of stuff.

3) Alternatively, a catalyst will allow the reaction to work at a much lower temperature. That reduces the energy used up in the reaction (the energy cost), which is good for sustainable development (see page 76) and can save a lot of money too.

4) There are disadvantages to using catalysts, though.

5) They can be very expensive to buy, and often need to be removed from the product and cleaned. They never get used up in the reaction though, so once you've got them you can use them over and over again.

6) Different reactions use different catalysts, so if you make more than one product at your plant, you'll probably need to buy different catalysts for them.

7) Catalysts can be 'poisoned' by impurities, so they stop working, e.g. sulfur impurities can poison the iron catalyst used in the Haber process (used to make ammonia for fertilisers). That means you have to keep your reaction mixture very clean.

Catalysts are like great jokes — they can be used over and over...

And they're not only used in industry... every useful chemical reaction in the human body is catalysed by a biological catalyst (an enzyme). If the reactions in the body were just left to their own devices, they'd take so long to happen, we couldn't exist. Quite handy then, these catalysts.

Energy Transfer in Reactions

Whenever chemical reactions occur <u>energy</u> is <u>transferred to</u> or <u>from</u> the <u>surroundings</u>.

In an <u>Exothermic</u> <u>Reaction, Heat is</u> <u>Given Out</u>

> An <u>EXOTHERMIC</u> <u>reaction</u> is one which <u>transfers energy</u> to the surroundings, usually in the form of <u>heat</u> and usually shown by a <u>rise in temperature.</u>

1) The best example of an <u>exothermic</u> reaction is <u>burning fuels</u> — also called <u>COMBUSTION</u>. This gives out a lot of heat — it's very exothermic.

2) <u>Neutralisation reactions</u> (acid + alkali) are also exothermic — see page 79.

3) Many <u>oxidation reactions</u> are exothermic. For example, adding sodium to water <u>produces heat</u>, so it must be <u>exothermic</u> — see page 61. The sodium emits <u>heat</u> and moves about on the surface of the water as it is oxidised.

4) Exothermic reactions have lots of <u>everyday uses</u>. For example, some <u>hand warmers</u> use the exothermic <u>oxidation of iron</u> in air (with a salt solution catalyst) to generate <u>heat</u>. <u>Self heating cans</u> of hot chocolate and coffee also rely on exothermic reactions between <u>chemicals</u> in their bases.

In an <u>Endothermic</u> <u>Reaction, Heat is</u> <u>Taken In</u>

> An <u>ENDOTHERMIC</u> <u>reaction</u> is one which <u>takes in energy</u> from the surroundings, usually in the form of <u>heat</u> and is usually shown by a <u>fall in temperature.</u>

Endothermic reactions are much <u>less common</u>. <u>Thermal decompositions</u> are a good example:

> Heat must be supplied to make calcium carbonate <u>decompose</u> to make calcium oxide.
> $$CaCO_3 \rightarrow CaO + CO_2$$

Endothermic reactions also have everyday uses. For example, some <u>sports injury packs</u> use endothermic reactions — they <u>take in heat</u> and the pack becomes very <u>cold</u>. More <u>convenient</u> than carrying ice around.

Reversible Reactions <u>Can Be</u> Endothermic <u>and</u> Exothermic

In reversible reactions (see page 76), if the reaction is <u>endothermic</u> in <u>one direction</u>, it will be <u>exothermic</u> in the <u>other direction</u>. The <u>energy absorbed</u> by the endothermic reaction is <u>equal</u> to the <u>energy released</u> during the exothermic reaction. A good example is the <u>thermal decomposition of hydrated copper sulfate</u>.

endothermic

hydrated copper sulfate ⇌ anhydrous copper sulfate + water

exothermic

"Anhydrous" just means "without water", and "hydrated" means "with water".

1) If you <u>heat blue hydrated</u> copper(II) sulfate crystals it drives the water off and leaves <u>white anhydrous</u> copper(II) sulfate powder. This is endothermic. 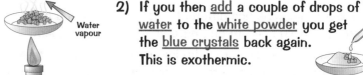 Water vapour

2) If you then <u>add</u> a couple of drops of <u>water</u> to the <u>white powder</u> you get the <u>blue crystals</u> back again. This is exothermic.

<u>Right, so burning gives out heat — really...</u>

This whole energy transfer thing is a fairly simple idea — don't be put off by the long words. Remember, "<u>exo-</u>" = <u>exit</u>, "<u>-thermic</u>" = <u>heat</u>, so an exothermic reaction is one that <u>gives out</u> heat. And "<u>endo-</u>" = erm... the other one. Okay, so there's no easy way to remember that one. Tough.

Energy

Whenever chemical reactions occur, there are changes in <u>energy</u>. This means that when chemicals get together, things either hot up or cool right off. I'll give you a heads up — this page is a good 'un.

Energy Must Always be Supplied _to Break Bonds..._
...and Energy is Always Released _When_ Bonds Form

1) During a chemical reaction, <u>old bonds</u> are <u>broken</u> and <u>new bonds</u> are <u>formed</u>.

2) Energy must be <u>supplied</u> to break <u>existing bonds</u> — so bond breaking is an <u>endothermic</u> process. Energy is <u>released</u> when new bonds are <u>formed</u> — so bond formation is an <u>exothermic</u> process.

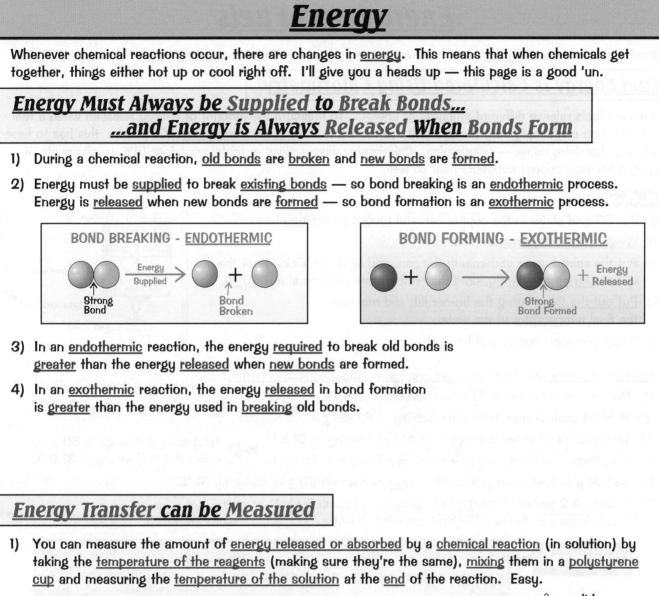

BOND BREAKING - <u>ENDOTHERMIC</u>

Energy Supplied

Strong Bond

Bond Broken

BOND FORMING - <u>EXOTHERMIC</u>

Energy Released

Strong Bond Formed

3) In an <u>endothermic</u> reaction, the energy <u>required</u> to break old bonds is <u>greater</u> than the energy <u>released</u> when <u>new bonds</u> are formed.

4) In an <u>exothermic</u> reaction, the energy <u>released</u> in bond formation is <u>greater</u> than the energy used in <u>breaking</u> old bonds.

Energy Transfer can be _Measured_

1) You can measure the amount of <u>energy released or absorbed</u> by a <u>chemical reaction</u> (in solution) by taking the <u>temperature of the reagents</u> (making sure they're the same), <u>mixing</u> them in a <u>polystyrene cup</u> and measuring the <u>temperature of the solution</u> at the <u>end</u> of the reaction. Easy.

2) The biggest <u>problem</u> with energy measurements is the amount of energy <u>lost to the surroundings</u>.

3) You can reduce it a bit by putting the polystyrene cup into a <u>beaker of cotton wool</u> to give <u>more insulation</u>, and putting a <u>lid</u> on the cup to reduce energy lost by <u>evaporation</u>.

4) This method works for reactions of <u>solids with water</u> (e.g. dissolving ammonium nitrate in water) as well as for <u>neutralisation</u> reactions.

thermometer

lid

polystyrene cup

reaction mixture

cotton wool

> Example:
> 1) Place 25 cm³ of dilute hydrochloric acid in a polystyrene cup, and record its temperature.
> 2) Put 25 cm³ of dilute sodium hydroxide in a measuring cylinder and record its temperature.
> 3) As long as they're at the same temperature, add the alkali to the acid and stir.
> 4) Take the temperature of the mixture every 30 seconds, and record the highest temperature it reaches.

Save energy — break fewer bonds...

So now you know why reactions are exothermic or endothermic — it's all about the <u>energy</u> needed to <u>make</u> or <u>break bonds</u>. Pretty easy I'd say — endothermic reactions take in heat because there is more energy needed to make new bonds than to break old ones. And exothermic reactions do the opposite. Super. Splendid.

Energy and Fuels

Burning <u>fuels</u> releases <u>energy</u>. Just how much energy you can find using <u>calorimetry</u>. Bet you can't wait.

Fuel Energy is Calculated Using Calorimetry

Different fuels release <u>different amounts of energy</u>. To measure the amount of energy released when a fuel is burnt, you can simply burn the fuel and use the flame to <u>heat up some water</u>. Of course, this has to have a fancy chemistry name — <u>calorimetry</u>. Calorimetry uses a <u>glass</u> or <u>metal container</u> (it's usually made of <u>copper</u> because copper conducts heat so well).

Method:

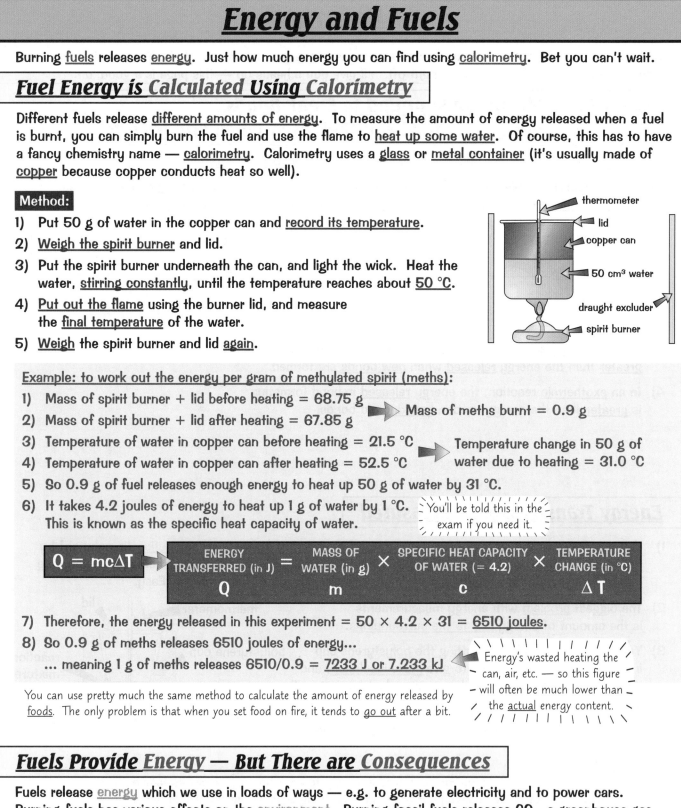

1) Put 50 g of water in the copper can and <u>record its temperature</u>.
2) <u>Weigh the spirit burner</u> and lid.
3) Put the spirit burner underneath the can, and light the wick. Heat the water, <u>stirring constantly</u>, until the temperature reaches about <u>50 °C</u>.
4) <u>Put out the flame</u> using the burner lid, and measure the <u>final temperature</u> of the water.
5) <u>Weigh</u> the spirit burner and lid <u>again</u>.

thermometer
lid
copper can
50 cm³ water
draught excluder
spirit burner

Example: to work out the energy per gram of methylated spirit (meths):

1) Mass of spirit burner + lid before heating = 68.75 g
2) Mass of spirit burner + lid after heating = 67.85 g
    ➡ Mass of meths burnt = 0.9 g
3) Temperature of water in copper can before heating = 21.5 °C
4) Temperature of water in copper can after heating = 52.5 °C
    ➡ Temperature change in 50 g of water due to heating = 31.0 °C
5) So 0.9 g of fuel releases enough energy to heat up 50 g of water by 31 °C.
6) It takes 4.2 joules of energy to heat up 1 g of water by 1 °C. This is known as the specific heat capacity of water.

You'll be told this in the exam if you need it.

$$Q = mc\Delta T$$

ENERGY TRANSFERRED (in J)	=	MASS OF WATER (in g)	×	SPECIFIC HEAT CAPACITY OF WATER (= 4.2)	×	TEMPERATURE CHANGE (in °C)
Q		m		c		ΔT

7) Therefore, the energy released in this experiment = 50 × 4.2 × 31 = <u>6510 joules</u>.
8) So 0.9 g of meths releases 6510 joules of energy...
   ... meaning 1 g of meths releases 6510/0.9 = <u>7233 J or 7.233 kJ</u>

Energy's wasted heating the can, air, etc. — so this figure will often be much lower than the actual energy content.

You can use pretty much the same method to calculate the amount of energy released by <u>foods</u>. The only problem is that when you set food on fire, it tends to <u>go out</u> after a bit.

Fuels Provide Energy — But There are Consequences

Fuels release <u>energy</u> which we use in loads of ways — e.g. to generate electricity and to power cars. Burning fuels has various effects on the <u>environment</u>. Burning fossil fuels releases CO_2, a greenhouse gas. This causes <u>global warming</u> and other types of <u>climate change</u>. It'll be <u>expensive</u> to slow down these effects, and to put things right. Developing alternative energy sources (e.g. tidal power) costs money.

Crude oil is <u>running out</u>. We use <u>a lot of fuels</u> made from crude oil (e.g. <u>petrol and diesel</u>) and as it runs out it will get more expensive. This means that everything that's <u>transported</u> by lorry, train or plane gets more expensive too. So the <u>price of crude oil</u> has a big economic effect.

Energy from fuels — it's a burning issue...

Alrighty. A bit of <u>method</u>, a few <u>sums</u> and some <u>social 'n' environmental gubbins</u> to round it off. A useful thing to remember is that energy values can be measured in <u>calories</u> instead of joules (1 calorie = 4.2 joules). Oh, and being familiar with the equation Q = mcΔT may come in quite handy as well.

Bond Energies

This is about calculating the stuff that you found by experiment on the previous page.

Energy Level Diagrams Show if it's Exo- or Endo-thermic

In exothermic reactions ΔH is –ve ← ΔH is the energy change.

1) This shows an exothermic reaction — the products are at a lower energy than the reactants. The difference in height represents the energy given out in the reaction (per mole). ΔH is –ve here.

2) The initial rise in the line represents the energy needed to break the old bonds. This is the activation energy.

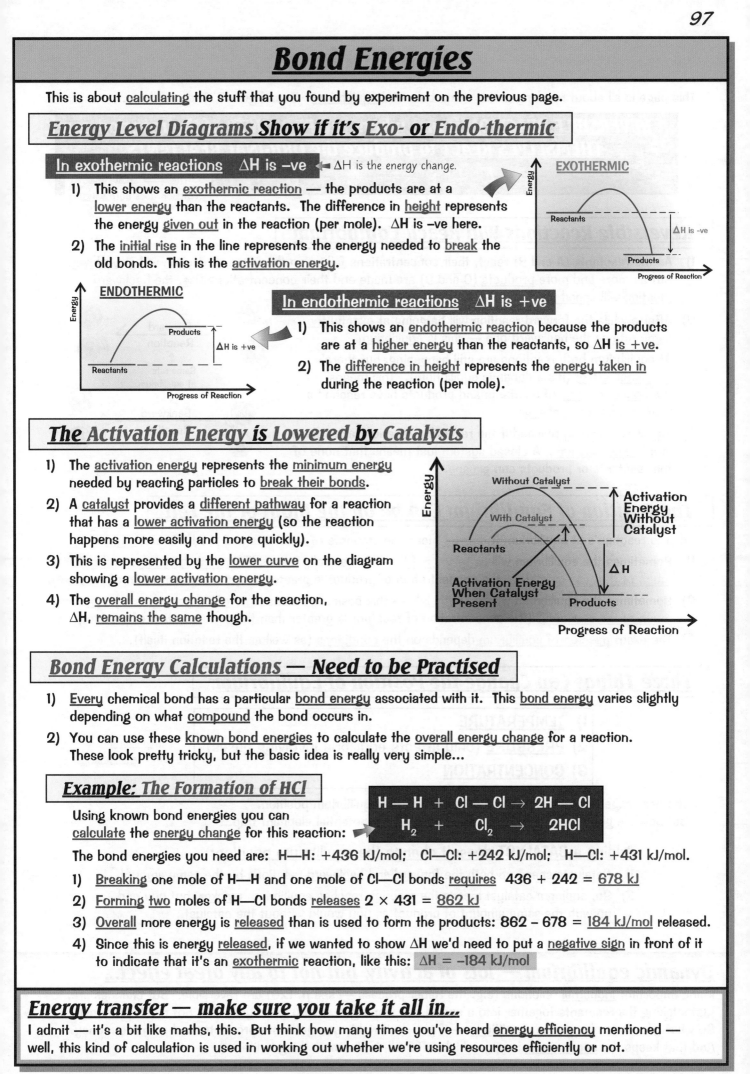

In endothermic reactions ΔH is +ve

1) This shows an endothermic reaction because the products are at a higher energy than the reactants, so ΔH is +ve.

2) The difference in height represents the energy taken in during the reaction (per mole).

The Activation Energy is Lowered by Catalysts

1) The activation energy represents the minimum energy needed by reacting particles to break their bonds.

2) A catalyst provides a different pathway for a reaction that has a lower activation energy (so the reaction happens more easily and more quickly).

3) This is represented by the lower curve on the diagram showing a lower activation energy.

4) The overall energy change for the reaction, ΔH, remains the same though.

Bond Energy Calculations — Need to be Practised

1) Every chemical bond has a particular bond energy associated with it. This bond energy varies slightly depending on what compound the bond occurs in.

2) You can use these known bond energies to calculate the overall energy change for a reaction. These look pretty tricky, but the basic idea is really very simple...

Example: The Formation of HCl

Using known bond energies you can calculate the energy change for this reaction:

$$H—H + Cl—Cl \rightarrow 2H—Cl$$
$$H_2 + Cl_2 \rightarrow 2HCl$$

The bond energies you need are: H—H: +436 kJ/mol; Cl—Cl: +242 kJ/mol; H—Cl: +431 kJ/mol.

1) Breaking one mole of H—H and one mole of Cl—Cl bonds requires 436 + 242 = 678 kJ

2) Forming two moles of H—Cl bonds releases 2 × 431 = 862 kJ

3) Overall more energy is released than is used to form the products: 862 – 678 = 184 kJ/mol released.

4) Since this is energy released, if we wanted to show ΔH we'd need to put a negative sign in front of it to indicate that it's an exothermic reaction, like this: ΔH = –184 kJ/mol

Energy transfer — make sure you take it all in...

I admit — it's a bit like maths, this. But think how many times you've heard energy efficiency mentioned — well, this kind of calculation is used in working out whether we're using resources efficiently or not.

Equilibrium

This page is all about reaching a nice cosy, balanced <u>equilibrium</u>. Sound good? Then read on...

> ## A <u>REVERSIBLE REACTION</u> is one where the <u>PRODUCTS</u> of the reaction can <u>THEMSELVES REACT</u> to produce the <u>ORIGINAL REACTANTS</u>
>
> $$A + B \rightleftharpoons C + D$$

The '$\rightleftharpoons$' shows the reaction goes <u>both ways</u>.

Reversible Reactions <u>Will Reach</u> Equilibrium

1) As the <u>reactants</u> (A and B) react, their concentrations <u>fall</u> — so the <u>forward reaction</u> will <u>slow down</u>. But as more and more <u>products</u> (C and D) are made and their concentrations <u>rise</u>, the <u>backward reaction</u> will <u>speed up</u>.

2) After a while the forward reaction will be going at <u>exactly the same rate</u> as the backward one — this is <u>equilibrium</u>.

3) At equilibrium <u>both</u> reactions are still <u>happening</u>, but there's <u>no overall effect</u> (it's a dynamic equilibrium). This means the <u>concentrations</u> of reactants and products have reached a balance and <u>won't change</u>.

4) Equilibrium is only reached if the reversible reaction takes place in a '<u>closed system</u>'. A <u>closed system</u> just means that none of the reactants or products can <u>escape</u>.

Forward Reaction

Same rate at equilibrium

Backward Reaction

<u>The Position of Equilibrium</u> <u>Can be on the</u> <u>Right</u> <u>or the Left</u>

When a reaction's at equilibrium it <u>doesn't</u> mean the amounts of reactants and products are <u>equal</u>.

1) Sometimes the equilibrium will <u>lie to the right</u> — this basically means "<u>lots of the products and not much of the reactants</u>" (i.e. the concentration of product is greater than the concentration of reactant).

2) Sometimes the equilibrium will <u>lie to the left</u> — this basically means "<u>lots of the reactants but not much of the products</u>" (the concentration of reactant is greater than the concentration of product).

3) The exact <u>position of equilibrium</u> depends on the <u>conditions</u> (as well as the reaction itself).

Three Things <u>Can Change</u> the Position of Equilibrium:

> 1) <u>TEMPERATURE</u>
> 2) <u>PRESSURE</u> (only affects equilibria involving gases)
> 3) <u>CONCENTRATION</u>

1 <u>equilibrium</u>, but 2 <u>equilibria</u>.

The next page tells you <u>why</u> these things affect the equilibrium position.
But now's a good time to make a mental note of this potential elephant trap...

> ### Adding a <u>CATALYST</u> doesn't change the equilibrium position:
> 1) Catalysts speed up <u>both</u> the <u>forward</u> and <u>backward</u> reactions by the <u>same amount</u>.
> 2) So, adding a catalyst means the reaction reaches equilibrium <u>quicker</u>, but you end up with the <u>same amount</u> of product as you would without the catalyst.

<u>Dynamic equilibrium — lots of activity, but not to any great effect...</u>*

Many important <u>industrial</u> reactions (e.g. the Haber process — see p. 126) are reversible. But chances are, just sticking the reactants together into a sealed box won't give a very good <u>yield</u> (i.e. not much product). So what you do is change the <u>conditions</u> — if you do it right, you get more products, and so more money. And that keeps the <u>accountants</u> happy, which, after all, is the main thing in life.

* Much like the England football team.

Changing Equilibrium

Now here's an interesting thing — if you <u>change</u> the conditions, the equilibrium will try to <u>counteract</u> that change. So if you <u>decrease</u> the <u>temperature</u>, the equilibrium will move to <u>produce more heat</u>. Sneaky.

The Equilibrium Tries to Minimise Any Changes You Make

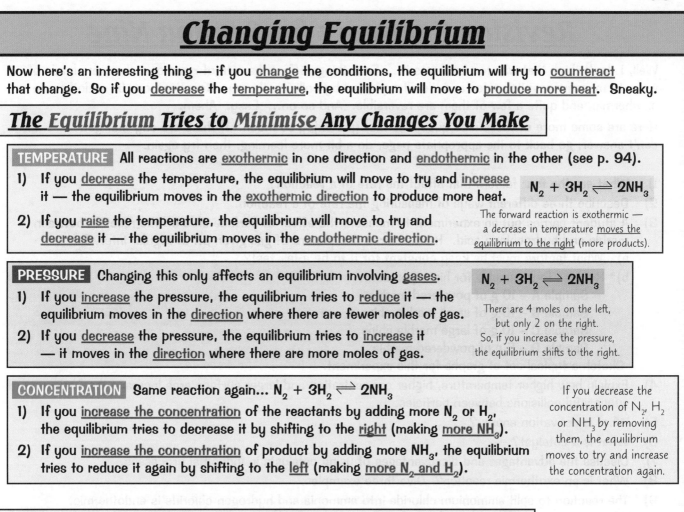

TEMPERATURE All reactions are <u>exothermic</u> in one direction and <u>endothermic</u> in the other (see p. 94).

1) If you <u>decrease</u> the temperature, the equilibrium will move to try and <u>increase</u> it — the equilibrium moves in the <u>exothermic direction</u> to produce more heat.

$N_2 + 3H_2 \rightleftharpoons 2NH_3$

2) If you <u>raise</u> the temperature, the equilibrium will move to try and <u>decrease</u> it — the equilibrium moves in the <u>endothermic direction</u>.

The forward reaction is exothermic — a decrease in temperature <u>moves the equilibrium to the right</u> (more products).

PRESSURE Changing this only affects an equilibrium involving <u>gases</u>.

$N_2 + 3H_2 \rightleftharpoons 2NH_3$

1) If you <u>increase</u> the pressure, the equilibrium tries to <u>reduce</u> it — the equilibrium moves in the <u>direction</u> where there are fewer moles of gas.

2) If you <u>decrease</u> the pressure, the equilibrium tries to <u>increase</u> it — it moves in the <u>direction</u> where there are more moles of gas.

There are 4 moles on the left, but only 2 on the right. So, if you increase the pressure, the equilibrium shifts to the right.

CONCENTRATION Same reaction again... $N_2 + 3H_2 \rightleftharpoons 2NH_3$

1) If you <u>increase the concentration</u> of the reactants by adding more N_2 or H_2, the equilibrium tries to decrease it by shifting to the <u>right</u> (making <u>more NH_3</u>).

2) If you <u>increase the concentration</u> of product by adding more NH_3, the equilibrium tries to reduce it again by shifting to the <u>left</u> (making <u>more N_2 and H_2</u>).

If you decrease the concentration of N_2, H_2 or NH_3 by removing them, the equilibrium moves to try and increase the concentration again.

Make Sure You Can Read Equilibrium Tables and Graphs

You might be asked to <u>interpret data</u> about <u>equilibrium</u>, so you'd better know what you're doing. The Haber process (see page 126) is a great example of all this...

$$N_2 + 3H_2 \rightleftharpoons 2NH_3$$

The forward reaction is exothermic.

First off, a table...

Pressure (atmospheres)	100	200	300	400	500
% of ammonia in reaction mixture at 450 °C	14	26	34	39	42

1) As the <u>pressure increases</u>, the proportion of ammonia <u>increases</u> (exactly what you'd expect — since increasing the pressure shifts the equilibrium to the side with fewer moles of gas — here, the right).

And now a graph...

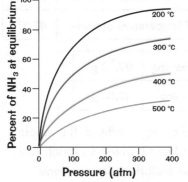

2) This time, each different line represents a different temperature.

3) As the temperature <u>increases</u>, the proportion of ammonia <u>decreases</u> (the backward reaction is endothermic, so this speeds up to try and reduce the temperature again).

4) The conditions that will give you <u>most ammonia</u> are <u>high pressure</u> and <u>low temperature</u>.

An equilibrium is like a particularly stubborn mule...

It's good science this stuff. <u>You</u> do one thing, and the <u>reaction</u> does the other. On the face of it, that sounds like it'd be pretty <u>annoying</u>, but in reality it's what gives you <u>control</u> of what happens. And in <u>industry</u>, control is what makes the whole shebang profitable. Mmmm... Money.

Revision Summary for Section Nine

Well, I don't think that was too bad, was it... Four things affect the rate of reactions, there are loads of ways to measure reaction rates and it's all explained by collision theory. Reactions can be endothermic or exothermic, and quite a few of them are reversible. And so on... Easy. Ahem.

Here are some more of those nice questions that you enjoy so much. If there are any that you can't answer, go back to the appropriate page, do a bit more learning, then try again.

1) What are the four factors that affect the rate of a reaction?
2) Describe three different ways of measuring the rate of a reaction.
3) A student carries out an experiment to measure the effect of surface area on the reaction between marble and hydrochloric acid. He measures the amount of gas given off at regular intervals.
   a) What factors must he keep constant for it to be a fair test?
   b)* He uses four samples for his experiment:
      Sample A – 10 g of powdered marble
      Sample B – 10 g of small marble chips
      Sample C – 10 g of large marble chips
      Sample D – 5 g of powdered marble
   Sketch a typical set of graphs for this experiment.
4) Explain how higher temperature, higher concentration and larger surface area increase the frequency of successful collisions between particles.
5) What is activation energy?
6) What is a catalyst?
7) Discuss the advantages and disadvantages of using catalysts in industrial processes.
8) What is an exothermic reaction? Give three examples.
9) The reaction to split ammonium chloride into ammonia and hydrogen chloride is endothermic. What can you say for certain about the reverse reaction?
10) An acid and an alkali were mixed in a polystyrene cup, as shown to the right. The acid and alkali were each at 20 °C before they were mixed. After they were mixed, the temperature of the solution reached 24 °C.
    a) State whether this reaction is exothermic or endothermic.
    b) Explain why the cotton wool is used.

20 cm³ of dilute sulfuric acid + 20 cm³ of dilute sodium hydroxide solution

cotton wool

11) Is energy released when bonds are formed or when bonds are broken?
12) The apparatus below is used to measure how much energy is released when pentane is burnt. It takes 4.2 joules of energy to heat 1 g of water by 1 °C.
    a)*Using the following data, and the equation Q = mc ΔT, calculate the amount of energy per gram of pentane.

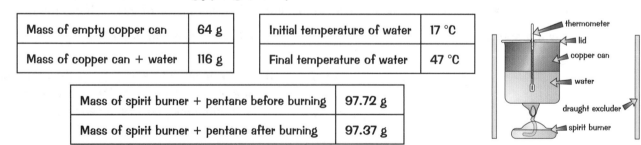

Mass of empty copper can	64 g
Mass of copper can + water	116 g

Initial temperature of water	17 °C
Final temperature of water	47 °C

Mass of spirit burner + pentane before burning	97.72 g
Mass of spirit burner + pentane after burning	97.37 g

thermometer
lid
copper can
water
draught excluder
spirit burner

   b) A data book says that pentane has 49 kJ/g of energy. Why is the amount you calculated different?
13) Explain why the price of bananas might rise if we keep burning so much fuel.
14) a) Draw energy level diagrams for exothermic and endothermic reactions.
    b) Explain how bond breaking and forming relate to these diagrams.
15) What is a reversible reaction? Explain why it could reach an equilibrium.
16) Describe how three different factors affect the position of equilibrium.

* Answers on page 140.

Section Nine — Reaction Rates and Energy Changes

Analytical Procedures

The next few pages are all about finding out exactly what is contained in a mystery substance.

Qualitative *Analysis Tells You What a Sample Contains*

1) Qualitative analysis tells you which substances are present in a sample.

2) It doesn't tell you how much of each substance there is — that's where quantitative analysis comes in.

Quantitative *Analysis Tells You How Much it Contains*

1) Quantitative analysis tells you how much of a substance is present in a sample.

2) It can be used to work out the molecular formula of the sample. E.g. if you had a sample containing carbon and hydrogen you'd know it was a hydrocarbon, but without quantitative analysis, you won't know if it's methane, butane or even 3,4-dimethylheptane...

Chemical Analysis *is Carried Out On Samples*

1) You usually analyse just a sample of the material under test. There's quite a few reasons for this.

2) It might be very difficult to test all of the material if you've got an awful lot of it — or you might want to just test a small bit so that you can use the rest for something else.

3) Taking a sample also means that if something goes wrong with the test, you can go back for another sample and try again.

4) A sample must represent the bulk of the material being tested — it wouldn't tell you anything very useful if it didn't.

Samples *are Analysed in Solution*

Samples are usually tested in solution. A solution is made by dissolving the sample in a solvent. There are two types of solution — aqueous and non-aqueous. Which type of solution you use depends on the type of substance you're testing.

An aqueous solution means the solvent is water. They're shown by the state symbol (aq).	A non-aqueous solution means the solvent is anything other than water — e.g. ethanol.

Standard Procedures *Mean Everyone Does Things the Same Way*

Whether testing chemicals or measuring giraffes, scientists follow 'standard procedures' — clear instructions describing exactly how to carry out these practical tasks.

1) Standard procedures are agreed methods of working — they are chosen because they're the safest, most effective and most accurate methods to use.

2) Standard procedures can be agreed within a company, nationally, or internationally.

3) They're useful because wherever and whenever a test is done, the result should always be the same — it should give reliable results every time.

4) There are standard procedures for the collection and storage of a sample, as well as how it should be analysed.

Analysis — don't they do that on Match of the Day?...

If you're trying to detect a certain substance in a sample, then there has to be a reasonable amount of the stuff you're looking for. You'd stand no chance of finding one molecule in a great big cake.

Tests for Positive Ions

Say you've got a compound, but you <u>don't know</u> what it is. Well, you'd want to <u>identify</u> it... that's only natural. And that's what the next couple of pages are all about. Tests for <u>positive ions</u> first...

Flame Tests *Identify Metal Ions*

Compounds of some metals burn with a characteristic colour.

Remember, metals always form positive ions.

1) You can test for various <u>metal ions</u> by putting your substance in a <u>flame</u> and seeing what <u>colour</u> the flame goes.

<u>Lithium</u>, Li^+, gives a crimson flame.
<u>Sodium</u>, Na^+, gives a yellow flame.
<u>Potassium</u>, K^+, gives a lilac flame.

<u>Calcium</u>, Ca^{2+}, gives a red flame.
<u>Barium</u>, Ba^{2+}, gives a green flame.

2) To flame-test a compound in the lab, dip a <u>clean wire loop</u> into a sample of the compound, and put the wire loop in the clear blue part of the Bunsen flame (the hottest bit). First make sure the wire loop is really clean by dipping it into <u>hydrochloric acid</u> and rinsing it with <u>distilled water</u>.

Some Metal Ions *Form a Coloured Precipitate* with NaOH

This is also a test for metal ions, but it's slightly more involved. Concentrate now...

1) Many <u>metal hydroxides</u> are <u>insoluble</u> and precipitate out of solution when formed. Some of these hydroxides have a <u>characteristic colour</u>.

2) So in this test you add a few drops of <u>sodium hydroxide</u> solution to a solution of your mystery compound — all in the hope of forming an insoluble hydroxide.

3) If you get a <u>coloured insoluble hydroxide</u> you can then tell which metal was in the compound.

"Metal"	Colour of precipitate	Ionic Equation
Calcium, Ca^{2+}	White	$Ca^{2+}_{(aq)} + 2OH^-_{(aq)} \rightarrow Ca(OH)_{2(s)}$
Copper(II), Cu^{2+}	Blue	$Cu^{2+}_{(aq)} + 2OH^-_{(aq)} \rightarrow Cu(OH)_{2(s)}$
Iron(II), Fe^{2+}	Green	$Fe^{2+}_{(aq)} + 2OH^-_{(aq)} \rightarrow Fe(OH)_{2(s)}$
Iron(III), Fe^{3+}	Brown	$Fe^{3+}_{(aq)} + 3OH^-_{(aq)} \rightarrow Fe(OH)_{3(s)}$
Aluminium, Al^{3+}	White at first. But then redissolves in excess NaOH to form a colourless solution.	$Al^{3+}_{(aq)} + 3OH^-_{(aq)} \rightarrow Al(OH)_{3(s)}$ then $Al(OH)_{3(s)} + OH^-_{(aq)} \rightarrow Al(OH)_4^-_{(aq)}$
Magnesium, Mg^{2+}	White	$Mg^{2+}_{(aq)} + 2OH^-_{(aq)} \rightarrow Mg(OH)_{2(s)}$

Ionic Equations *Show Just the Useful Bits* of Reactions

1) The reactions in the above table are <u>ionic equations</u>. Ionic equations are 'half' a full equation, if you like. For example: $Ca^{2+}_{(aq)} + 2OH^-_{(aq)} \longrightarrow Ca(OH)_{2(s)}$

2) This <u>just</u> shows the formation of <u>calcium hydroxide</u> from the <u>calcium ions</u> and the <u>hydroxide ions</u>. The <u>full</u> equation in the above reaction would be (if you started off with <u>calcium chloride</u>, say):

$$CaCl_{2(aq)} + 2NaOH_{(aq)} \longrightarrow Ca(OH)_{2(s)} + 2NaCl_{(aq)}$$

3) But the formation of <u>sodium chloride</u> is of no great interest here — it's not helping <u>identify</u> the compound.

4) So the ionic equation just concentrates on the <u>good bits</u>.

It isn't any old ion — get a positive identification...

Just think of an ionic equation as a bit like Match of the Day — an <u>edited highlights package</u>.

Section Ten — Chemical Analysis and Electrolysis

Tests for Negative Ions

So now maybe you know what the <u>positive</u> part of your mystery substance is (see previous page). Now it's time to test for the <u>negative</u> bit.

Testing for Carbonates — Check for CO₂

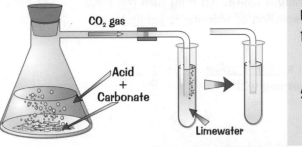

CO₂ gas

Acid + Carbonate

Limewater

First things first — the test for carbon dioxide (CO_2).

1) You can test to see if a gas is <u>carbon dioxide</u> by bubbling it through <u>limewater</u>. If it is <u>carbon dioxide</u>, the <u>limewater turns cloudy</u>.

2) You can use this to test for <u>carbonate</u> ions (CO_3^{2-}), since carbonates react with <u>dilute acids</u> to form <u>carbon dioxide</u>.

Acid + Carbonate → Salt + Water + Carbon dioxide

Tests for Halides and Sulfates

You can test for certain ions by seeing if a <u>precipitate</u> is formed after these reactions...

Halide Ions

To test for <u>chloride</u> (Cl⁻), <u>bromide</u> (Br⁻) or <u>iodide</u> (I⁻) ions, add <u>dilute nitric acid</u> (HNO_3), followed by <u>silver nitrate solution</u> ($AgNO_3$).

A <u>chloride</u> gives a white precipitate of <u>silver chloride</u>.

$$Ag^+_{(aq)} + Cl^-_{(aq)} \longrightarrow AgCl_{(s)}$$

A <u>bromide</u> gives a cream precipitate of <u>silver bromide</u>.

$$Ag^+_{(aq)} + Br^-_{(aq)} \longrightarrow AgBr_{(s)}$$

An <u>iodide</u> gives a yellow precipitate of <u>silver iodide</u>.

$$Ag^+_{(aq)} + I^-_{(aq)} \longrightarrow AgI_{(s)}$$

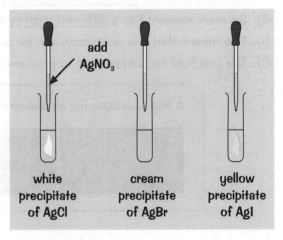

add AgNO₃

white precipitate of AgCl   cream precipitate of AgBr   yellow precipitate of AgI

Sulfate Ions

1) To test for a <u>sulfate</u> ion (SO_4^{2-}), <u>add dilute HCl</u>, followed by <u>barium chloride solution</u>, $BaCl_2$.

2) A <u>white</u> precipitate of <u>barium sulfate</u> means the original compound was a sulfate.

$$Ba^{2+}_{(aq)} + SO_4^{2-}_{(aq)} \longrightarrow BaSO_{4(s)}$$

Sherlock never looked so good in a lab coat...

So you might have to do loads of different chemical tests to find out all the information about your mystery substance. It's a bit like detective work — eliminating suspects, narrowing down possibilities, and so on. It's the kind of stuff <u>exam questions</u> are sometimes made of, by the way, so be warned. They might give you the <u>results</u> from several chemical tests, and you have to say what the substance is.

Section Ten — Chemical Analysis and Electrolysis

Gas Tests and Spectroscopy

There are the old-fashioned ways of identifying substances, and then there are the new-fangled ones...

You can Test for Chlorine and Hydrogen Gases

Chlorine Chlorine bleaches damp litmus paper, turning it white. (It may turn red for a moment first though — that's because a solution of chlorine is acidic.)

Damp litmus paper

Hydrogen Hydrogen makes a "squeaky pop" with a lighted splint. (The noise comes from the hydrogen burning with the oxygen in the air to form H_2O.)

Squeaky pop!

H_2 gas

Oxygen You can test for oxygen by checking if the gas relights a glowing splint.

Glowing splint

oxygen

Each Element Gives a Characteristic Line Spectrum

1) When heated, the electrons in an atom are excited, and release energy as light.
2) The wavelengths emitted can be recorded as a line spectrum.
3) Different elements emit different wavelengths of light. This is due to each element having a different electron arrangement (see page 12).
4) So each element has a different pattern of wavelengths, and a different line spectrum.
5) This means that line spectrums can be used to identify elements.
6) The practical technique used to produce line spectrums is called spectroscopy.

A line spectrum for an element will look something like this:

Line Spectrums Have Identified New Elements

New practical techniques (e.g. spectroscopy) have allowed scientists to discover new elements. Some of these elements simply wouldn't have been discovered without the development of these techniques.

- Caesium and rubidium were both discovered by their line spectrum.
- Helium was discovered in the line spectrum of the Sun.

Spectroscopy — it's a flaming useful technique...

There are quite nifty ways of identifying elements as well as things like your bog standard glowing splint test. Some of these methods are so nifty that they've actually allowed us to discover elements that otherwise might still be unknown to us. I think that deserves your attention... If nothing else, think of all the pretty colours.

Analysis — Chromatography

There are several different types of <u>chromatography</u> — it's a cracking analytical method.

Chromatography <u>uses</u> Two Phases

<u>Chromatography</u> is an analytical method used to <u>separate</u> the substances in a mixture. You can then use it to <u>identify</u> the substances. There are lots of different <u>types</u> of chromatography — but they all have two 'phases':

- A <u>mobile phase</u> — where the molecules <u>can</u> move. This is always a <u>liquid</u> or a <u>gas</u>.
- A <u>stationary phase</u> — where the molecules <u>can't</u> move. This can be a <u>solid</u> or a really <u>thick liquid</u>.

1) The components in the mixture <u>separate</u> out as the mobile phase moves across the stationary phase.
2) How quickly a chemical <u>moves</u> depends on how it "<u>distributes</u>" itself between the two phases — this is why <u>different chemicals</u> separate out and end up at <u>different points</u> (see below).
3) The molecules of each chemical constantly <u>move</u> between the mobile and the stationary phases.
4) They are said to reach a "<u>dynamic equilibrium</u>" — at equilibrium the amount leaving the stationary phase for the <u>mobile phase</u> is the <u>same</u> as the amount leaving the mobile phase for the <u>stationary phase</u>. But be careful, this doesn't (necessarily) mean there is the <u>same</u> amount of chemical in each phase.

Don't panic — this will all make more sense once you've read the rest of the page...

<u>In</u> Paper Chromatography <u>the Stationary Phase is</u> Paper

1) In paper chromatography, a <u>spot</u> of the substance being tested is put onto a <u>baseline</u> on the paper.
2) The bottom of the paper is placed in a beaker containing a <u>solvent</u>, such as ethanol or water. The solvent is the <u>mobile phase</u>.
3) The <u>stationary phase</u> is the <u>chromatography paper</u> (often <u>filter paper</u>).

spots of chemicals

baseline

sample

solvent →

Here's what happens:

> 1) The <u>solvent moves up</u> the paper.
> 2) The chemicals in the sample <u>dissolve</u> in the solvent and move between it and the paper. This sets up an equilibrium between the solvent and the paper.
> 3) When they're in the <u>mobile phase</u> the chemicals <u>move</u> up the paper with the solvent.
> 4) Before the solvent reaches the <u>top</u> of the paper, the paper is <u>removed</u> from the beaker.
> 5) The different chemicals in the sample form <u>separate spots</u> on the paper. The chemicals that <u>spend more time</u> in the <u>mobile phase</u> than the <u>stationary phase</u> form spots <u>further up</u> the paper.

The amount of time the molecules spend in each <u>phase</u> depends on two things:

> 1) How <u>soluble</u> they are in the solvent.
> 2) How <u>attracted</u> they are to the paper.

So molecules with a <u>higher solubility</u> in the solvent, and which are <u>less attracted</u> to the paper, will spend <u>more time in the mobile phase</u> — and they'll be carried further up the paper.

Thin-Layer Chromatography <u>has a Different</u> Stationary Phase

1) <u>Thin-layer chromatography</u> (TLC) is very similar to paper chromatography, but the <u>stationary phase</u> is a thin layer of solid — e.g. <u>silica gel</u> spread onto a glass plate.
2) The <u>mobile phase</u> is a solvent such as ethanol (just like in paper chromatography).

<u>Learning about this — it's just a phase you go through...</u>

The tricky thing about understanding how chromatography works is that you <u>can't</u> see the chemicals moving between the two phases — you'll just have to believe that it <u>does</u> happen.

Analysis — Chromatography

You can Calculate the R_f Value for Each Chemical

1) The result of chromatography analysis is called a <u>chromatogram</u>.

2) Some of the spots on the chromatogram might be <u>colourless</u>. If they are, you need to use a <u>locating agent</u> to show where they are, e.g. you might have to spray the chromatogram with a reagent.

3) You can work out the <u>R_f values</u> for <u>spots</u> (solutes) on a chromatogram. An R_f value is the <u>ratio</u> between the distance travelled by the dissolved substance (the solute) and the distance travelled by the solvent. You can find them using the formula:

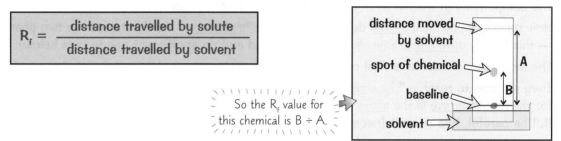

$$R_f = \frac{\text{distance travelled by solute}}{\text{distance travelled by solvent}}$$

So the R_f value for this chemical is B ÷ A.

distance moved by solvent

spot of chemical

A

baseline

B

solvent

4) Chromatography is often carried out to see if a certain substance is present in a mixture. You run a <u>pure, known sample</u> of the substance alongside the unknown mixture. If the R_f values match, the substances may be the <u>same</u> (although it doesn't definitely prove they are the same).

5) Chemists use substances called <u>standard reference materials</u> (SRMs) to check the identities of substances. These have carefully <u>controlled concentrations and purities</u>.

Gas Chromatography is a Bit More High-Tech

<u>Gas chromatography</u> (GC) is used to analyse <u>unknown substances</u> too. If they're not already gases, then they have to be vaporised.

- The <u>mobile phase</u> is an <u>unreactive gas</u> such as nitrogen.
- The <u>stationary phase</u> is a <u>viscous</u> (thick) <u>liquid</u>, such as an oil.

The process is quite <u>different</u> from paper chromatography and TLC:

1) The unknown mixture is <u>injected</u> into a long tube <u>coated</u> on the inside with the <u>stationary phase</u>.

2) The mixture <u>moves</u> along the tube with the <u>mobile phase</u> until it <u>comes out</u> the other end. Like in the other chromatography methods, the substances are <u>distributed</u> between the phases.

3) The <u>time</u> it takes a chemical to <u>travel through</u> the tube is called the <u>retention time</u>.

4) The retention time is <u>different</u> for each chemical — it's what's used to <u>identify</u> it.

mobile phase enters here

sample injected

detector

tube coated with stationary phase

temperature controlled oven

The <u>chromatogram</u> from gas chromatography is a graph. Each <u>peak</u> on the graph represents a <u>different chemical</u>.

- The <u>distance</u> along the x-axis is the <u>retention time</u> — which can be <u>looked up</u> to find out <u>what</u> the chemical is.
- The peak <u>height</u> shows you <u>how much</u> of that chemical was in the sample.

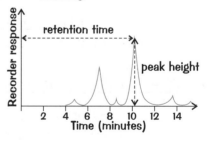

Recorder response

retention time

peak height

Time (minutes)

Comb-atography — identifies mysterious things in your hair...

Chromatography works by showing how mystery chemicals get <u>distributed</u> between mobile and stationary phases — that's what the <u>R_f value</u> represents. All chemicals get distributed differently, so that's how you can tell which is which. It's great — all you need is some paper and a bit of solvent.

Moles and Titration

There's a nice little experiment you can do to find out how much alkali you need to <u>neutralise</u> an acid. As I'm feeling generous, there's also some useful stuff about <u>moles</u> and <u>concentration</u> thrown in on this page too — you never know when it might come in handy...

"THE MOLE" is Simply the Name Given to a Certain Number

Just like "<u>a million</u>" is this many: 1 000 000;   or "<u>a billion</u>" is this many: 1 000 000 000, "<u>a mole</u>" is this many: 602 300 000 000 000 000 000 000   or 6.023×10^{23}.

1) And that's all it is. <u>Just a number</u>. The burning question, of course, is why is it such a silly long one like that, and with a six at the front?

2) The answer is that when you get <u>precisely that number</u> of atoms of <u>carbon-12</u> it weighs exactly <u>12 g</u>. So, get that number of atoms or molecules, <u>of any element or compound</u>, and conveniently, they <u>weigh</u> exactly the same number of <u>grams</u> as the relative atomic mass, A_r (or M_r) of the element or compound. This is arranged <u>on purpose</u> of course, to make things easier.

3) So, you can use <u>moles</u> as a <u>unit</u> of measurement when you're talking about an amount of a substance.

Concentration is a Measure of How Crowded Things Are

The <u>concentration</u> of a solution can be measured in <u>moles per dm³</u> (i.e. <u>moles per litre</u>). So 1 mole of stuff in 1 dm³ of solution has a concentration of <u>1 mole per dm³</u> (or 1 mol/dm³).

> The <u>more solute</u> you dissolve in a given volume, the <u>more crowded</u> the solute molecules are and the <u>more concentrated</u> the solution.

Concentration can also be measured in <u>grams per dm³</u>. So 56 grams of stuff dissolved in 1 dm³ of solution has a concentration of <u>56 grams per dm³</u>.

> 1 litre
> = 1000 cm³
> = 1 dm³

Titrations are Used to Find Out Concentrations

1) Titrations also allow you to find out <u>exactly</u> how much acid is needed to <u>neutralise</u> a quantity of alkali (or vice versa).

2) You put some <u>alkali</u> in a flask, along with some <u>indicator</u> — <u>phenolphthalein</u> or <u>methyl orange</u>. You don't use Universal indicator as it changes colour gradually — and you want a <u>definite</u> colour change.

3) Add the <u>acid</u>, a bit at a time, to the alkali using a <u>burette</u> — giving the flask a regular <u>swirl</u>. Go especially <u>slowly</u> (a drop at a time) when you think the alkali's almost neutralised.

4) The indicator <u>changes colour</u> when <u>all</u> the alkali has been <u>neutralised</u> — phenolphthalein is <u>pink</u> in <u>alkalis</u> but <u>colourless</u> in <u>acids</u>, and methyl orange is <u>yellow</u> in <u>alkalis</u> but <u>red</u> in <u>acids</u>.

5) <u>Record</u> the amount of acid used to <u>neutralise</u> the alkali. It's best to <u>repeat</u> this process a few times, making sure you get (pretty much) the same answer each time.

6) You can then take the <u>mean</u> of your results.

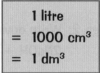

You can also do titrations the other way round — adding alkali to acid.

burette containing acid

These marks down the side show the volume of acid used.

alkali and indicator

If you can spell phenolphthalein, you deserve a GCSE...

There's a little bit at the end of the titration method that's pretty vital when you do <u>any</u> experiment — the bit about <u>repeating</u> the process to check your results. It's all to do with making sure your results are <u>reliable</u>. If you get the same result a number of times, you can have more faith in it than if it's a one-off.

Titration Calculations

I expect you're wondering what you can do with the results from a titration experiment (who wouldn't be). Well, you'll be relieved to know that they can be used to <u>calculate concentrations</u> of acids or alkalis.

You Can Calculate Concentration in Moles or in Grams

If you have the results of a titration experiment you can calculate the concentration of the acid when you know the concentration of the alkali (or vice versa).

Example 1: If you want the concentration in MOLES per dm³

Say you start off with <u>25 cm³</u> of sodium hydroxide in your flask, and you know that its concentration is <u>0.1 moles per dm³</u>.

You then find from your titration that it takes <u>30 cm³</u> of sulfuric acid (whose concentration you don't know) to neutralise the sodium hydroxide.

You can work out the <u>concentration</u> of the acid in <u>moles per dm³</u>.

> Concentration = moles ÷ volume, so you can make a handy formula triangle.
>
> Concentration (in mol/dm³)   Number of moles
> $$\frac{n}{c \times V}$$
> Volume (in dm³) One dm³ is a litre
>
> <u>Cover up</u> the thing you're trying to find — then what's left is the formula you need to use.

<u>Step 1</u>: Work out how many <u>moles</u> of the "known" substance you have using this formula:

Number of moles = concentration × volume
= 0.1 mol/dm³ × (25 / 1000) dm³ = <u>0.0025 moles of NaOH</u>

> Remember: 1000 cm³ = 1 dm³

> Use the formula triangle if it helps.

<u>Step 2</u>: Write down the <u>balanced equation</u> of the reaction...

$$2NaOH + H_2SO_4 \longrightarrow Na_2SO_4 + 2H_2O$$

...and work out how many <u>moles</u> of the "<u>unknown</u>" stuff you must have had.

Using the equation, you can see that for every <u>two moles</u> of sodium hydroxide you had...
...there was just <u>one mole</u> of sulfuric acid.
So if you had <u>0.0025 moles</u> of sodium hydroxide...
...you must have had 0.0025 ÷ 2 = <u>0.00125 moles of sulfuric acid</u>.

<u>Step 3</u>: Work out the concentration of the "unknown" stuff.
Concentration = number of moles ÷ volume
= 0.00125 mol ÷ (30 / 1000) dm³ = 0.041666... mol/dm³
= <u>0.0417 mol/dm³</u>

> Don't forget to put the units.

Example 2: If you want the concentration in GRAMS per dm³

They might ask you to find out the acid concentration in <u>grams per cubic decimetre</u> (<u>grams per litre</u>). If they do, don't panic — you just need another formula triangle.

<u>Step 1</u>: Work out the <u>relative formula mass</u> for the acid (you should be given the relative atomic masses, e.g. H = 1, S = 32, O = 16):

So, $H_2SO_4 = (1 \times 2) + 32 + (16 \times 4) = 98$

<u>Step 2</u>: Convert the concentration in <u>moles</u> (that you've already worked out) into concentration in <u>grams</u>. So, in 1 dm³:

> Use non-rounded answers in workings.

Mass in grams = moles × relative formula mass
= 0.041666... × 98 = 4.08333... g
So the <u>concentration in g/dm³ = 4.08 g/dm³</u>

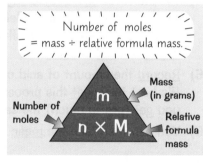

> Number of moles = mass ÷ relative formula mass.
>
> Mass (in grams)
> $$\frac{m}{n \times M_r}$$
> Number of moles   Relative formula mass

Need practice, you do — mmmm...

Scary. But if you get enough practice at these questions, then the fear will evaporate and you can tackle them with a smile on your face and a spring in your step. Remember, don't be baffled by <u>dm³</u> — it's just an overly complicated way of saying "<u>litre</u>", that's all. So "moles per dm³" means "<u>moles per litre</u>". Simple.

Purity

It's all very well making a product, but how do you know if it's any good or not...

Some Products Need to be Very Pure

1) The purity of a product will improve as it's being isolated. But to get a really pure product earlier stages (such as filtration, evaporation and crystallisation) will often need to be repeated.

2) Purifying and measuring the purity of a product are particularly important steps in the chemical industry. For example, in the production of...

> ...pharmaceuticals — it's really important to ensure drugs intended for human consumption are free from impurities — these could do more harm than good.
>
> ...petrochemicals — if there are impurities or contaminants in petrol products they could cause damage to a car's engine.

Titrations Can be Used to Measure the Purity of a Substance

The purity of a compound can be calculated using a titration:

Example — Determining the Purity of Aspirin

Say you start off with 0.2 g of impure aspirin dissolved in 25 cm³ of ethanol in your conical flask. You find from your titration that it takes 9.5 cm³ of 4 g/dm³ NaOH to neutralise the aspirin solution.

Step 1: Work out the concentration of the aspirin solution.
Here's the formula — just stick the numbers in the right places.

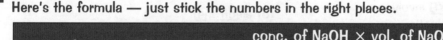

$$\text{conc. of aspirin solution} = 4.5 \times \frac{\text{conc. of NaOH} \times \text{vol. of NaOH}}{\text{vol. of aspirin solution}}$$

This number will change depending on which acid and alkali are used.

$$= 4.5 \times \frac{4 \times (9.5 / 1000)}{25 / 1000} = \underline{6.84 \text{ g/dm}^3}$$

The volumes need to be in dm³, not cm³, so you need to divide by 1000 to convert it.

1 dm³ = 1000 cm³

Step 2: Work out the mass of the aspirin. You'll need to use another formula:

$$\text{mass} = \text{concentration} \times \text{volume}$$

$$= 6.84 \times 25 / 1000 = \underline{0.171 \text{ g}}$$

This is the concentration of the aspirin solution.

In the 0.2 g that you started with, 0.171 g was aspirin.

Step 3: Calculate the purity, using the formula:

$$\% \text{ purity} = \frac{\text{calculated mass of substance}}{\text{mass of impure substance at start}} \times 100$$

$$= 0.171 \div 0.2 \times 100 = \underline{85.5\%}$$

This is the percentage purity of the aspirin.

This is the mass of aspirin that reacted in the titration.

This is the mass of impure aspirin you started with.

Concentrate when doing titrations...

Drugs need to be free of harmful impurities — but not all impurities are a problem. Some impurities are deliberately added to drugs like aspirin to make them taste better and bind together into a nice tablet shape.

Electrolysis

Hmm, electrolysis. A not-very-catchy title for quite a <u>sparky</u> subject...

Electrolysis Means "Splitting Up with Electricity"

1) If you pass an <u>electric current</u> through an <u>ionic substance</u> that's <u>molten</u> or in <u>solution</u>, it breaks down into the <u>elements</u> it's made of. This is called <u>electrolysis</u>.

2) It requires a <u>liquid</u> to <u>conduct</u> the <u>electricity</u>, called the <u>electrolyte</u>.

3) Electrolytes contain <u>free ions</u> — they're usually the <u>molten</u> or <u>dissolved ionic substance</u>.

4) In either case it's the <u>free ions</u> which <u>conduct</u> the electricity and allow the whole thing to work.

5) For an electrical circuit to be complete, there's got to be a <u>flow of electrons</u>. Electrons are taken <u>away from</u> ions at the <u>positive electrode</u> and <u>given to</u> other ions at the <u>negative electrode</u>. As ions gain or lose electrons they become atoms or molecules and are released.

NaCl dissolved

Molten NaCl

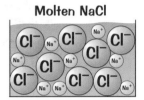

Electrolysis Reactions Involve Oxidation and Reduction

1) <u>Oxidation</u> is a <u>gain of oxygen</u> or a <u>loss of electrons</u> (see page 85).

2) <u>Reduction</u> is a <u>loss of oxygen</u> or a <u>gain of electrons</u>.

3) Electrolysis <u>ALWAYS</u> involves an oxidation and a reduction.

Oxidation	Reduction
Is	Is
Loss	Gain

(of <u>electrons</u>)

~ Remember it ~
~ as OIL RIG. ~

The Electrolysis of Molten Lead Bromide

When a salt (e.g. lead bromide) is molten it will conduct electricity.

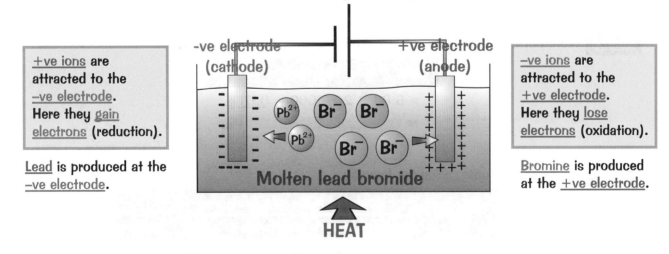

<u>+ve ions</u> are attracted to the <u>–ve electrode</u>. Here they <u>gain electrons</u> (reduction).

<u>Lead</u> is produced at the <u>–ve electrode</u>.

-ve electrode (cathode)

+ve electrode (anode)

Molten lead bromide

HEAT

<u>–ve ions</u> are attracted to the <u>+ve electrode</u>. Here they <u>lose electrons</u> (oxidation).

<u>Bromine</u> is produced at the <u>+ve electrode</u>.

1) At the <u>–ve electrode</u>, one lead ion <u>accepts</u> two electrons to become <u>one lead atom</u>.

2) At the <u>+ve electrode</u>, two bromide ions <u>lose</u> one electron each and become <u>one bromine molecule</u>.

Faster shopping at Tesco — use Electrolleys...

<u>Reduction</u> happens at the <u>negative</u> electrode and <u>oxidation</u> happens at the <u>positive</u>. There's no two ways about it. Electrolysis is used lots in <u>real life</u>, and it's nice to know how these things work, I reckon.

Electrolysis of Sodium Chloride Solution

As well as <u>molten substances</u> you can also electrolyse <u>solutions</u>. But first, a bit more about the <u>products</u>...

Reactivity Affects the Products Formed By Electrolysis

1) Sometimes there are <u>more than two free ions</u> in the electrolyte.
For example, if a salt is <u>dissolved in water</u> there will also be some <u>H$^+$</u> and <u>OH$^-$</u> ions.

2) At the <u>negative electrode</u>, if <u>metal ions</u> and <u>H$^+$ ions</u> are present, the metal ions will <u>stay in solution</u> if the metal is <u>more reactive</u> than hydrogen. This is because the more reactive an element, the keener it is to stay as ions. So, <u>hydrogen</u> will be produced unless the metal is <u>less reactive</u> than it.

3) At the <u>positive electrode</u>, if <u>OH$^-$</u> and <u>halide ions</u> (Cl$^-$, Br$^-$, I$^-$) are present then molecules of chlorine, bromine or iodine will be formed. If <u>no halide</u> is present, then <u>oxygen</u> will be formed.

The Electrolysis of Sodium Chloride Solution

When common salt (sodium chloride) is dissolved in water and electrolysed,
it produces three useful products — <u>hydrogen</u>, <u>chlorine</u> and <u>sodium hydroxide</u>.

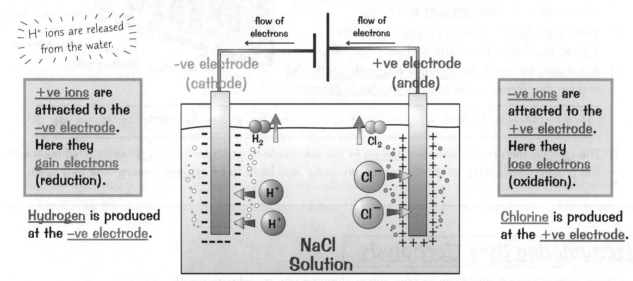

~ H$^+$ ions are released ~
~ from the water. ~

+ve ions are
attracted to the
−ve electrode.
Here they
gain electrons
(reduction).

Hydrogen is produced
at the −ve electrode.

flow of electrons

flow of electrons

−ve electrode
(cathode)

+ve electrode
(anode)

H$_2$

Cl$_2$

Cl$^-$

Cl$^-$

H$^+$

H$^+$

NaCl
Solution

−ve ions are
attracted to the
+ve electrode.
Here they
lose electrons
(oxidation).

Chlorine is produced
at the +ve electrode.

1) At the <u>negative electrode</u>, two hydrogen ions accept two electrons to become <u>one hydrogen molecule</u>.

2) At the <u>positive electrode</u>, two chloride (Cl$^-$) ions lose their electrons and become <u>one chlorine molecule</u>.

3) The <u>sodium ions</u> stay in solution because they're <u>more reactive</u> than hydrogen. <u>Hydroxide ions</u> from water are also left behind. This means that <u>sodium hydroxide</u> (NaOH) is left in the solution.

The Half-Equations — Make Sure the Electrons Balance

Half equations show the reactions at the electrodes. The main thing is to make sure the <u>number of electrons</u> is the <u>same</u> for <u>both half-equations</u>. For the electrolysis of sodium chloride the half-equations are:

~ You need to make ~
~ sure the atoms ~
~ are balanced too. ~

<u>Negative Electrode:</u> $2H^+ + 2e^- \rightarrow H_2$

<u>Positive Electrode:</u> $2Cl^- \rightarrow Cl_2 + 2e^-$

or $2Cl^- - 2e^- \rightarrow Cl_2$

For the electrolysis of molten lead bromide (previous page) the half equations would be:
$$Pb^{2+} + 2e^- \rightarrow Pb$$
and $2Br^- \rightarrow Br_2 + 2e^-$

Useful Products from the Electrolysis of Sodium Chloride Solution

The products of the electrolysis of sodium chloride solution are pretty useful in <u>industry</u>.

1) Chlorine has many uses, e.g. in the production of <u>bleach</u> and <u>plastics</u>.

2) Sodium hydroxide is a very strong <u>alkali</u> and is used <u>widely</u> in the <u>chemical industry</u>, e.g. to make <u>soap</u>.

SOAP

Extraction of Aluminium and Electroplating

I bet you never thought you'd see so many pages about electrolysis — well, you have now...

Electrolysis *is Used to Remove* Aluminium *from Its* Ore

1) Aluminium's a very <u>abundant</u> metal, but it is always found naturally in <u>compounds</u>.

2) Its main ore is <u>bauxite</u>, and after mining and purifying, a <u>white powder</u> is left.

3) This is <u>pure</u> aluminium oxide, Al_2O_3.

4) The <u>aluminium</u> has to be extracted from this using <u>electrolysis</u>.

Cryolite *is Used to* Lower *the* Temperature *(and Costs)*

1) Al_2O_3 has a very <u>high melting point</u> of over <u>2000 °C</u> — so melting it would be very <u>expensive</u>.

2) <u>Instead</u> the aluminium oxide is <u>dissolved</u> in <u>molten cryolite</u> (a less common ore of aluminium).

3) This brings the <u>temperature down</u> to about <u>900 °C</u>, which makes it much <u>cheaper</u> and <u>easier</u>.

4) The <u>electrodes</u> are made of <u>carbon</u> (graphite), a good conductor of electricity (see page 66).

5) <u>Aluminium</u> forms at the <u>negative electrode (cathode)</u> and <u>oxygen</u> forms at the <u>positive electrode (anode)</u>.

crust
carbon positive electrode (graphite)
carbon lining (graphite) for negative electrode
bauxite in molten cryolite
molten aluminium

> <u>Negative Electrode</u>: $Al^{3+} + 3e^- \rightarrow Al$ <u>Positive Electrode</u>: $2O^{2-} \rightarrow O_2 + 4e^-$

6) The <u>oxygen</u> then reacts with the <u>carbon</u> in the electrode to produce <u>carbon dioxide</u>. This means that the <u>positive electrodes</u> gradually get 'eaten away' and have to be <u>replaced</u> every now and again.

Electroplating *Uses Electrolysis*

1) Electroplating uses electrolysis to <u>coat</u> the <u>surface of one metal</u> with <u>another metal</u>, e.g. you might want to electroplate silver onto a brass cup to make it look nice.

2) The <u>negative electrode (cathode)</u> is the <u>metal object</u> you want to plate and the <u>positive electrode (anode)</u> is the <u>pure metal</u> you want it to be plated with. You also need the <u>electrolyte</u> to contain <u>ions</u> of the <u>plating metal</u>. (The ions that plate the metal object come from the solution, while the positive electrode keeps the solution 'topped up'.)

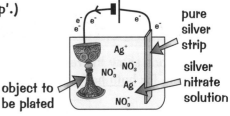

pure silver strip
silver nitrate solution
object to be plated

> <u>Example</u>: To electroplate <u>silver</u> onto a <u>brass cup</u>, you'd make the <u>brass cup</u> the negative electrode (to attract the positive silver ions), a lump of <u>pure silver</u> the positive electrode and dip them in a solution of <u>silver ions</u>, e.g. silver nitrate.

3) There are lots of different <u>uses</u> for electroplating:

• <u>Decoration</u>: <u>Silver</u> is <u>attractive</u>, but very <u>expensive</u>. It's much <u>cheaper</u> to plate a boring brass cup with silver, than it is to make the cup out of solid silver — but it looks just as <u>pretty</u>.

• <u>Conduction</u>: Metals like <u>copper</u> conduct <u>electricity</u> well — because of this they're often used to plate metals for <u>electronic circuits</u> and <u>computers</u>.

Silver electroplated text is worth a fortune...

From <u>extracting metals</u> to making them look all <u>shiny</u>, electrolysis has the answer. There are loads of metals you can use for electroplating — it's not just about copper and silver. The tricky bit is remembering that the metal <u>object you want to plate</u> is the <u>negative electrode</u> and the <u>metal</u> you're plating it with is the <u>positive electrode</u>.

Revision Summary for Section Ten

And that's it... the end of another section. Which means it's time for some more questions. There's no point in trying to duck out of these — and you know why... For these questions you'll need to know some common chemical tests. So my friend, if you find you don't know them, fear not... just look back in the book.

1) What is the difference between qualitative and quantitative analysis?

2) What is meant by the phrase 'standard procedures'? Why are they important?

3) Describe two ways of testing for metal ions.

4) How would you distinguish between solutions of:
   a) magnesium sulfate and aluminium sulfate,
   b) sodium bromide and sodium iodide,
   c) copper nitrate and copper sulfate?

5) How would you test for hydrogen gas?

6) Why does each element produce a different line spectrum?

7) What can line spectrums be used for?

8) Describe the two phases in chromatography.

9) What are the mobile and stationary phases in paper chromatography and thin-layer chromatography?

10)* What is the R_f value of a chemical that moves 4.5 cm when the solvent moves 12 cm?

11) What are the mobile and stationary phases in gas chromatography?

12) What is meant by 'retention time'?

13) What does the height of a peak on a gas chromatogram show?

14) Name a suitable indicator you could use in the titration of sulfuric acid and sodium hydroxide.

15)* In a titration, 49 cm³ of hydrochloric acid was required to neutralise 25 cm³ of sodium hydroxide with a concentration of 0.2 moles per dm³.
    Calculate the concentration of the hydrochloric acid in: a) mol/dm³   b) g/dm³

16) Why is purification of a product important?

17)* Calculate the purity of a 0.5 g aspirin tablet that contains 0.479 g of aspirin.

18) What is electrolysis? Explain why only liquids can be electrolysed.

19) Draw a detailed diagram with half equations showing the electrolysis of sodium chloride.

20) Give one industrial use of sodium hydroxide and two uses of chlorine.

21) Why is cryolite used during the electrolysis of aluminium oxide?

22) Give two different uses of electroplating.

* Answers on page 140.

Hardness of Water

Water where you live might be <u>hard</u> or <u>soft</u>. It depends on the <u>rocks</u> your water meets on its way to you.

Hard Water Makes Scum and Scale

1) With <u>soft water</u>, you get a nice <u>lather</u> with soap. But with <u>hard water</u> you get a <u>nasty scum</u> instead — unless you're using a soapless detergent. The problem is dissolved <u>calcium ions</u> and <u>magnesium ions</u> in the water (see below) reacting with the soap to make <u>scum</u> which is insoluble. So to get a decent lather you need to use <u>more soap</u> — and because soap <u>isn't free</u>, that means <u>more money</u> going down the drain.

2) When <u>heated</u>, hard water also forms furring or <u>scale</u> (mostly calcium carbonate) on the insides of pipes, boilers and kettles. Badly scaled-up pipes and boilers reduce the <u>efficiency</u> of heating systems, and may need to be <u>replaced</u> — all of which costs money. Scale can even <u>eventually block pipes</u>.

3) <u>Scale</u> is also a bit of a <u>thermal insulator</u>. This means that a <u>kettle</u> with scale on the <u>heating element</u> takes <u>longer to boil</u> than a <u>clean</u> non-scaled-up kettle — so it becomes <u>less efficient</u>.

Hardness is Caused by Ca^{2+} and Mg^{2+} Ions

1) Most hard water is hard because it contains lots of <u>calcium ions</u> and <u>magnesium ions</u>.

2) Rain falling on some types of rocks (e.g. <u>limestone</u>, <u>chalk</u> and <u>gypsum</u>) can dissolve compounds like <u>magnesium sulfate</u> (which is soluble), and <u>calcium sulfate</u> (which is also soluble, though only a bit).

Hard Water Isn't All Bad

1) Ca^{2+} ions are good for healthy <u>teeth</u> and <u>bones</u>.

2) Studies have found that people who live in <u>hard water</u> areas are at <u>less risk</u> of developing <u>heart disease</u> than people who live in soft water areas. This could be to do with the <u>minerals</u> in hard water.

Remove the Dissolved Ca^{2+} and Mg^{2+} Ions to Make Hard Water Soft

There are two kinds of hardness — <u>temporary</u> and <u>permanent</u>.
<u>Temporary hardness</u> is caused by the <u>hydrogencarbonate</u> ion, $\underline{HCO_3^-}$, in $Ca(HCO_3)_2$.
<u>Permanent hardness</u> is caused by dissolved <u>calcium sulfate</u> (among other things).

1) <u>Temporary hardness</u> is removed by <u>boiling</u>. When <u>heated</u>, the calcium hydrogencarbonate <u>decomposes</u> to form <u>calcium carbonate</u> which is <u>insoluble</u>. This solid is the 'limescale' on your kettle.

   e.g.
   $$\text{calcium hydrogencarbonate} \rightarrow \text{calcium carbonate} + \text{water} + \text{carbon dioxide}$$
   $$Ca(HCO_3)_{2(aq)} \rightarrow CaCO_{3(s)} + H_2O_{(l)} + CO_{2(g)}$$

   This <u>won't work</u> for permanent hardness, though. Heating a <u>sulfate</u> ion does <u>nowt</u>.

2) <u>Both types of hardness</u> can be softened by adding washing soda (<u>sodium carbonate</u>, Na_2CO_3) to it. The added <u>carbonate ions</u> react with the Ca^{2+} and Mg^{2+} ions to make an <u>insoluble precipitate</u> of calcium carbonate and magnesium carbonate. The Ca^{2+} and Mg^{2+} ions are no longer dissolved in the water so they can't make it hard.

   e.g. $$Ca^{2+}_{(aq)} + CO_3^{2-}_{(aq)} \rightarrow CaCO_{3(s)}$$

3) <u>Both types of hardness</u> can also be removed by running water through 'ion exchange columns' which are sold in shops. The columns have lots of <u>sodium ions</u> (or <u>hydrogen ions</u>) and 'exchange' them for calcium or magnesium ions in the water that runs through them.

   e.g. $$Na_2Resin_{(s)} + Ca^{2+}_{(aq)} \rightarrow CaResin_{(s)} + 2Na^+_{(aq)}$$

   ('Resin' is a huge insoluble resin molecule.)

And if the water's really hard, you can chip your teeth...

Hard water — good thing or bad thing... Well, it provides minerals that are good for health, but it creates an awful lot of <u>unnecessary expense</u>. All in all, it's a bit of a drag. Never mind — you win some and you lose some.

Hardness of Water

If you live in an area of <u>hard water</u> you might already know about it because you sometimes get a grimy layer of <u>scale</u> floating on the top of your tea and you get through more shampoo than you can shake a stick at...

<u>You Can Use</u> Titration <u>to Compare the Hardness of</u> Water Samples

Method

1) Fill a burette with <u>50 cm³ of soap solution</u>.
2) Add <u>50 cm³</u> of the first <u>water</u> sample into a flask.
3) Use the burette to add <u>1 cm³ of soap solution</u> to the flask.
4) Put a <u>bung</u> in the flask and <u>shake</u> for 10 seconds.
5) <u>Repeat</u> steps 3 and 4 until a <u>good lasting lather</u> is formed. (A lasting lather is one where the <u>bubbles cover the surface</u> for <u>at least 30 seconds</u>.)
6) <u>Record</u> how much soap was needed to create a lasting lather.
7) <u>Repeat</u> steps 1-6 with the other water samples.
8) Next, <u>boil fresh samples</u> of each type of water for <u>ten minutes</u>, and <u>repeat</u> the experiment.

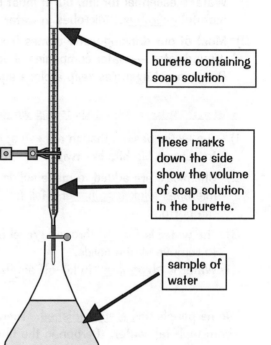

shake   shake   shake

Good lather

burette containing soap solution

These marks down the side show the volume of soap solution in the burette.

sample of water

Results

This method was carried out on <u>3 different samples of water</u> — <u>distilled</u> water, <u>local tap water</u> and <u>imported tap water</u>. Here's the <u>table of results</u>:

Sample	Volume of soap solution needed to give a good lather	
	using unboiled water in cm³	using boiled water in cm³
Distilled	1	1
Local water	7	1
Imported water	14	8

The results tell you the following things about the water:

1) Distilled water contains little or no <u>hardness</u> — only the <u>minimum</u> amount of soap was needed.
2) The sample of <u>imported water</u> contains <u>more hardness</u> than <u>local water</u> — <u>more soap</u> was needed to produce a lather.
3) The local water contains only <u>temporary hardness</u> — all the hardness is <u>removed by boiling</u>. You can tell because the same amount of soap was needed for <u>boiled local water</u> as for <u>distilled water</u>.
4) The imported water contains both <u>temporary</u> and <u>permanent hardness</u>. 8 cm³ of soap is still needed to produce a lather after boiling.
5) If your brain's really switched on, you'll see that the local water and the imported water contain the <u>same amount</u> of <u>temporary hardness</u>. In both cases, the amount of soap needed in the <u>boiled</u> sample is <u>6 cm³ less</u> than in the <u>unboiled</u> sample.

My water's harder than yours...

One thing that I've never understood is that they sell water softeners in areas that <u>already</u> have soft water. Hmm... Anyhow, the usual message here. There's nothing on this page which should surprise you too much. In fact, the only surprise is that people have put this much effort into testing water hardness in the first place.

Water Quality

It's easy to take water for granted... turn on the tap, and there it is — nice, clean water. The water you drink's been round the block a few times — so there's some fancy chemistry needed to make it drinkable.

Drinking Water Needs to Be Good Quality

1) Water's essential for life, but it must be free of poisonous salts (e.g. phosphates and nitrates) and harmful microbes. Microbes in water can cause diseases such as cholera and dysentery.

2) Most of our drinking water comes from reservoirs. Water flows into reservoirs from rivers and groundwater — water companies choose to build reservoirs where there's a good supply of clean water. Government agencies keep a close eye on pollution in reservoirs, rivers and groundwater.

Water from reservoirs goes to the water treatment works for treatment:

1) The water passes though a mesh screen to remove big bits like twigs.
2) Chemicals are added to make solids and microbes stick together and fall to the bottom.
3) The water is filtered through gravel beds to remove all the solids.
4) Water is chlorinated to kill off any harmful microbes left.

Water from reservoir → screening → removal of solids and microbes → filtration → chlorination → People's houses

Some people still aren't satisfied. They buy filters that contain carbon or silver to remove substances from their tap water. Carbon in the filters removes chlorine taste and silver is supposed to kill bugs. Some people in hard water areas buy water softeners which contain ion exchange resins (see p. 114).

Totally pure water with nothing dissolved in it can be produced by distillation — boiling water to make steam and condensing the steam. This process is too expensive to produce tap water — bags of energy would be needed to boil all the water we use. Distilled water is used in chemistry labs.

You'd use pure water to make a solution of (say) KBr, because you wouldn't want any other ions mucking it up.

Adding Fluoride and Chlorine to Water Has Disadvantages

1) Fluoride is added to drinking water in some parts of the country because it helps to reduce tooth decay. Chlorine is added to prevent disease (see above). So far so good. However...

2) Some studies have linked adding chlorine to water with an increase in certain cancers. Chlorine can react with other natural substances in water to produce toxic by-products which some people think could cause cancer.

3) In high doses fluoride can cause cancer and bone problems in humans, so some people believe that fluoride shouldn't be added to drinking water. There is also concern about whether it's right to 'mass medicate' — people can choose whether to use a fluoride toothpaste, but they can't choose whether their tap water has added fluoride.

4) Levels of chemicals added to drinking water need to be carefully monitored. For example, in some areas the water may already contain a lot of fluoride, so adding more could be harmful.

The water you drink has been through 7 people already...

Well, it's possible. It's also possible that the water you're drinking used to be part of the Atlantic Ocean. Or it could have been drunk by Alexander the Great. Or part of an Alpine glacier. Aye, it gets about a bit, does water. And remember... tap water isn't pure — but it's drinkable, and that's the main thing.

Detergents

Here's a splendid example of <u>better, cleaner living</u> through Chemistry.

Detergents **have a** Hydrophilic **Head and a** Hydrophobic **Tail**

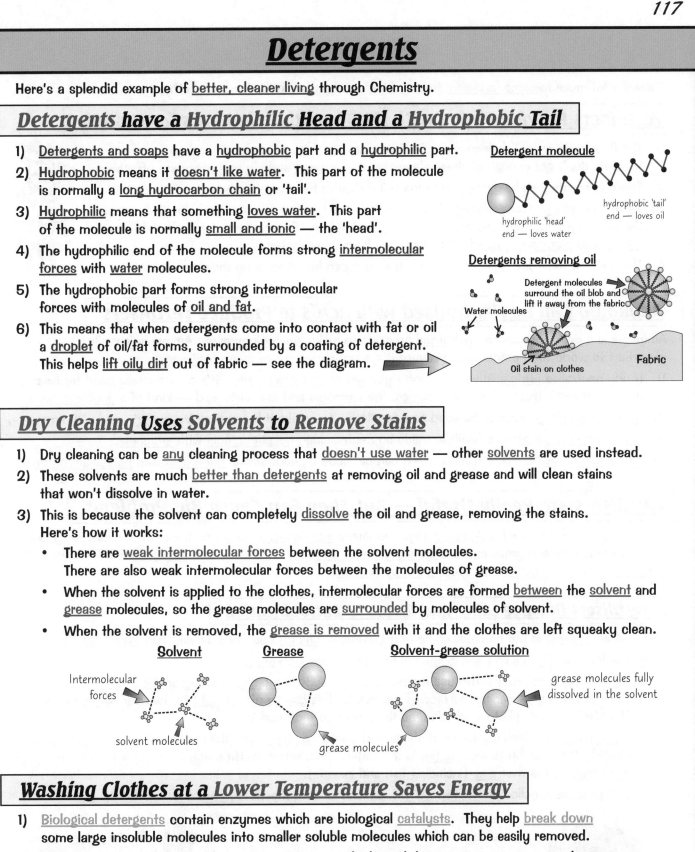

1) <u>Detergents and soaps</u> have a <u>hydrophobic</u> part and a <u>hydrophilic</u> part.

2) <u>Hydrophobic</u> means it <u>doesn't like water</u>. This part of the molecule is normally a <u>long hydrocarbon chain</u> or 'tail'.

3) <u>Hydrophilic</u> means that something <u>loves water</u>. This part of the molecule is normally <u>small and ionic</u> — the 'head'.

4) The hydrophilic end of the molecule forms strong <u>intermolecular forces</u> with <u>water</u> molecules.

5) The hydrophobic part forms strong intermolecular forces with molecules of <u>oil and fat</u>.

6) This means that when detergents come into contact with fat or oil a <u>droplet</u> of oil/fat forms, surrounded by a coating of detergent. This helps <u>lift oily dirt</u> out of fabric — see the diagram.

Detergent molecule

hydrophobic 'tail' end — loves oil

hydrophilic 'head' end — loves water

<u>Detergents removing oil</u>

Detergent molecules surround the oil blob and lift it away from the fabric

Water molecules

Oil stain on clothes

Fabric

Dry Cleaning **Uses** Solvents **to Remove Stains**

1) Dry cleaning can be <u>any</u> cleaning process that <u>doesn't use water</u> — other <u>solvents</u> are used instead.

2) These solvents are much <u>better than detergents</u> at removing oil and grease and will clean stains that won't dissolve in water.

3) This is because the solvent can completely <u>dissolve</u> the oil and grease, removing the stains. Here's how it works:

- There are <u>weak intermolecular forces</u> between the solvent molecules. There are also weak intermolecular forces between the molecules of grease.

- When the solvent is applied to the clothes, intermolecular forces are formed <u>between</u> the <u>solvent</u> and <u>grease</u> molecules, so the grease molecules are <u>surrounded</u> by molecules of solvent.

- When the solvent is removed, the <u>grease is removed</u> with it and the clothes are left squeaky clean.

Solvent

Intermolecular forces

solvent molecules

Grease

grease molecules

Solvent-grease solution

grease molecules fully dissolved in the solvent

Washing Clothes at a **Lower Temperature Saves Energy**

1) <u>Biological detergents</u> contain enzymes which are biological <u>catalysts</u>. They help <u>break down</u> some large insoluble molecules into smaller soluble molecules which can be easily removed.

2) Most enzymes work best at lower temperatures so biological detergents mean you can do your washing at <u>lower temperatures</u>. Turning your washing machine down from a 40 °C to a 30 °C wash uses about 40% <u>less energy</u>. And that means it saves you money on your bills... Hooray.

3) At higher temperatures, the enzymes found in biological detergents are <u>denatured</u> (destroyed). This means that biological detergents don't work as well at temperatures above about 40 °C.

4) Washing at a <u>cooler temperature</u> also means you can wash more <u>delicate clothes</u> in the washing machine.

Detergents and solvents — what a washout...

There's lots of clever stuff on this page — so no excuses for not knowing how <u>detergents</u> and <u>dry cleaning solvents</u> work. And remember that washing your clothes at a lower temperature can <u>save energy</u> and <u>money</u>.

Fertilisers

There's a lot more to using fertilisers than making your garden look nice and pretty...

Fertilisers Provide Plants with the Essential Elements for Growth

1) The three main essential elements in fertilisers are nitrogen, phosphorus and potassium. If plants don't get enough of these elements, their growth and life processes are affected.

2) These elements may be missing from the soil if they've been used up by a previous crop.

3) Fertilisers replace these missing elements or provide more of them. This helps to increase the crop yield, as the crops can grow faster and bigger. For example, fertilisers add more nitrogen to plant proteins, which makes the plants grow faster.

4) The fertiliser must first dissolve in water before it can be taken in by the crop roots.

Ammonia Can be Neutralised with Acids to Produce Fertilisers

Ammonia is a base and can be neutralised by acids to make ammonium salts. Ammonia is really important to world food production, because it's a key ingredient of many fertilisers.

1) If you neutralise nitric acid with ammonia you get ammonium nitrate. It's an especially good fertiliser because it has nitrogen from two sources, the ammonia and the nitric acid — kind of a double dose.

2) Ammonium sulfate can also be used as a fertiliser. You make it by neutralising sulfuric acid with ammonia.

3) Ammonium phosphate is a fertiliser made by neutralising phosphoric acid with ammonia.

4) Potassium nitrate is also a fertiliser — it can be made by neutralising nitric acid with potassium hydroxide.

Fertilisers are Really Useful — But They Can Cause Big Problems

The population of the world is rising rapidly. Fertilisers increase crop yield, so the more fertiliser we make, the more crops we can grow, and the more people we can feed. But if we use too many fertilisers we risk polluting our water supplies and causing eutrophication.

Fertilisers Damage Lakes and Rivers — Eutrophication

1) When fertiliser is put on fields some of it inevitably runs off and finds its way into rivers and streams.

2) The level of nitrates and phosphates in the river water increases.

3) Algae living in the river water use the nutrients to multiply rapidly, creating an algal bloom (a carpet of algae near the surface of the river). This blocks off the light to the river plants below. The plants cannot photosynthesise, so they have no food and they die.

4) Aerobic bacteria feed on the dead plants and start to multiply. As the bacteria multiply they use up all the oxygen in the water. As a result pretty much everything in the river dies (including fish and insects).

> Aerobic just means that they need oxygen to live.

5) This process is called EUTROPHICATION, which basically means 'too much of a good thing'.

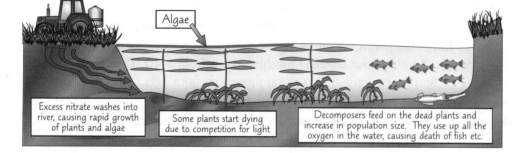

Excess nitrate washes into river, causing rapid growth of plants and algae

Algae

Some plants start dying due to competition for light

Decomposers feed on the dead plants and increase in population size. They use up all the oxygen in the water, causing death of fish etc.

As the picture shows, too many nitrates in the water cause a sequence of 'mega-growth', 'mega-death' and 'mega-decay' involving most of the plant and animal life in the water.

There's nowt wrong wi' just spreadin' muck on it...

Unfortunately, no matter how good something is, there's nearly always a downside. If you've ever seen a lake or river covered in a layer of green algae then it's pretty likely that it's caused by eutrophication. Hmmm.

Getting Energy From Hydrogen

Fuel cells are great — they use hydrogen and oxygen to make electricity.

Hydrogen *and* Oxygen *Give Out* Energy *When They React*

1) Hydrogen and oxygen react to produce water.

2) The reaction between hydrogen and oxygen is exothermic — it releases energy.

3) You can show this on an energy level diagram.

4) The higher a line is on the diagram the more energy the substances have. For example the H_2 and O_2 molecules have a higher energy than the H_2O molecules.

5) When the new bonds are formed the excess energy is given out in the form of heat.

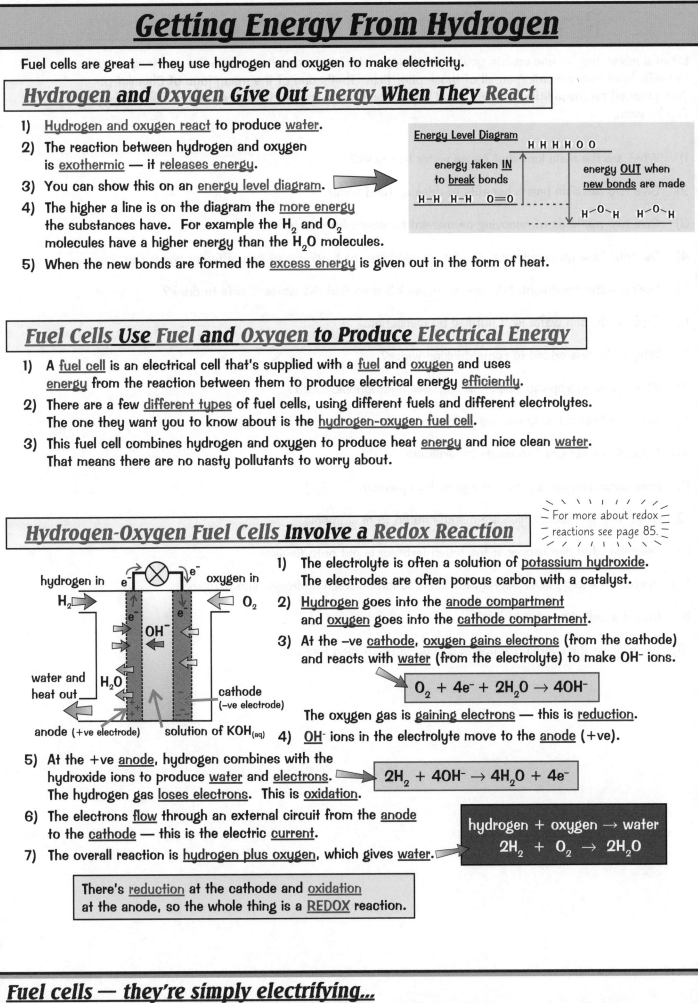

Energy Level Diagram

H H H H O O

energy taken IN to break bonds

H–H H–H O=O

energy OUT when new bonds are made

H–O–H H–O–H

Fuel Cells *Use* Fuel *and* Oxygen *to Produce* Electrical Energy

1) A fuel cell is an electrical cell that's supplied with a fuel and oxygen and uses energy from the reaction between them to produce electrical energy efficiently.

2) There are a few different types of fuel cells, using different fuels and different electrolytes. The one they want you to know about is the hydrogen-oxygen fuel cell.

3) This fuel cell combines hydrogen and oxygen to produce heat energy and nice clean water. That means there are no nasty pollutants to worry about.

Hydrogen-Oxygen Fuel Cells *Involve a Redox Reaction*

For more about redox reactions see page 85.

hydrogen in e^- e^- oxygen in

H_2 O_2

e^- OH⁻ e^-

water and heat out H_2O

cathode (–ve electrode)

anode (+ve electrode)   solution of KOH₍aq₎

1) The electrolyte is often a solution of potassium hydroxide. The electrodes are often porous carbon with a catalyst.

2) Hydrogen goes into the anode compartment and oxygen goes into the cathode compartment.

3) At the –ve cathode, oxygen gains electrons (from the cathode) and reacts with water (from the electrolyte) to make OH⁻ ions.

$$O_2 + 4e^- + 2H_2O \rightarrow 4OH^-$$

The oxygen gas is gaining electrons — this is reduction.

4) OH⁻ ions in the electrolyte move to the anode (+ve).

5) At the +ve anode, hydrogen combines with the hydroxide ions to produce water and electrons. The hydrogen gas loses electrons. This is oxidation.

$$2H_2 + 4OH^- \rightarrow 4H_2O + 4e^-$$

6) The electrons flow through an external circuit from the anode to the cathode — this is the electric current.

7) The overall reaction is hydrogen plus oxygen, which gives water.

hydrogen + oxygen → water
$$2H_2 + O_2 \rightarrow 2H_2O$$

There's reduction at the cathode and oxidation at the anode, so the whole thing is a REDOX reaction.

Fuel cells — they're simply electrifying...

This page is tough stuff. Close the book, scribble it all down and see what you know — it's the only way.

Revision Summary for Section Eleven

Bit of a mixed bag — one minute you're pondering water hardness, the next you're worrying about the effects of eutrophication on all of those poor fish. That's one of the many joys of Chemistry. Test yourself on these little beauties. And if you don't know any, flick back and learn the stuff again. Top banana.

1) What are the main ions that cause water hardness?

2) Give two possible health benefits of drinking hard water.

3) Give two methods of removing permanent hardness from water.

4) Describe how you could use titration to compare the hardness of two different water samples.

5) During water treatment, how are microbes killed so that the water is safe to drink?

6) Explain why tap water isn't purified by distillation.

7) Why is fluoride added to some drinking water?

8) What does hydrophilic mean? What does hydrophobic mean?

9) Describe how solvents remove stains.

10) Name three essential elements in fertilisers.

11) How does nitrogen increase the growth of plants?

12) Name two fertilisers which are manufactured from ammonia.

13) Describe what can happen if too much fertiliser is put onto fields.

14) Sketch an energy level diagram for the reaction between hydrogen and oxygen.

15) Give the definition of a fuel cell.

16) Write down the overall reaction in an hydrogen-oxygen fuel cell.

The Chemical Industry

Absolutely loads of some types of chemicals are used — such as <u>fertilisers</u>. Well they don't just grow on trees. No, they have to be <u>made</u>. And made they are — on a <u>massive scale</u>.

Some Chemicals are Produced on a Large Scale...

There are certain chemicals that industries need <u>thousands and thousands of tonnes</u> of every year — <u>ammonia</u>, <u>sulfuric acid</u>, <u>sodium hydroxide</u> and <u>phosphoric acid</u> are four examples. Chemicals that are produced on a <u>large scale</u> are called <u>bulk chemicals</u>.

...And Some are Produced on a Small Scale

Some chemicals aren't needed in such large amounts — but that doesn't mean they're any less important.

Chemicals produced on a <u>smaller scale</u> are called <u>fine chemicals</u>. Some examples are <u>drugs</u>, <u>food additives</u> and <u>fragrances</u>.

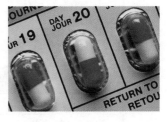

New Chemical Products Need Lots of Research

Before new chemical products are made, a huge amount of <u>research and development</u> work goes on. This can take <u>years</u>, and be <u>really expensive</u>, but it's worth it in the end if the company makes lots of <u>money</u> out of the new product.

For example, to make a new production process run efficiently a new <u>catalyst</u> might have to be found. This is likely to involve:

1) <u>Testing</u> potential catalysts using a process of <u>trial and error</u>.

2) Making <u>computer models</u> of the reaction to try to work out which substance might work as a catalyst.

3) <u>Designing or refining the manufacture</u> of the catalyst to make sure that the new product can be mass-produced <u>safely</u>, <u>efficiently</u>, and <u>cost effectively</u>.

4) Investigating the <u>risks to the environment</u> of using the new catalyst and trying to <u>minimise</u> them.

5) <u>Monitoring</u> the <u>quality</u> of the product to make sure that it is not affected by the catalyst.

These jobs, and lots of other types of work, are done by people in the <u>chemical industry</u>.

Government Regulations Protect People and the Environment

Governments place <u>strict controls</u> on everything to do with <u>chemical processes</u>. This is done to <u>protect workers</u>, the <u>general public</u> and the <u>environment</u>. For example, there are regulations about...

1) <u>Using chemicals</u> — e.g. <u>sulfuric acid</u> is sprayed on potato fields to <u>destroy</u> the leaves and stalks of the potato plants and make harvesting easier. Government <u>regulations</u> <u>restrict</u> how much acid can be used and require <u>signs</u> to be displayed to <u>warn</u> the public.

2) <u>Storage</u> — many <u>dangerous chemicals</u> have to be stored in locked storerooms. <u>Poisonous</u> chemicals must be stored in either <u>sealed containers</u> or well-ventilated store cupboards.

3) <u>Transport</u> — e.g. lorries transporting chemicals must display <u>hazard symbols</u> and <u>identification numbers</u> to help the emergency services deal <u>safely</u> with any accidents and spills.

Fine chemicals — by appointment to Her Majesty, The Queen...

Producing <u>bulk</u> chemicals is like painting a house — it's huge, so you slap that paint on with a <u>great big</u> brush. Producing <u>fine</u> chemicals is like painting a picture — much <u>smaller</u> and more <u>fiddly</u>. Got it?

Life Cycle Assessments

If a company wants to manufacture a new product, they carry out a life cycle assessment (LCA).
This looks at every stage of the product's life to assess the impact it would have on the environment.

Life Cycle Assessments Show Total Environmental Costs

A life cycle assessment (LCA) looks at each stage of the life of a product — from making the material from natural raw materials (see page 124), making the product from the material, using the product and disposing of the product. It works out the potential environmental impact of each stage.

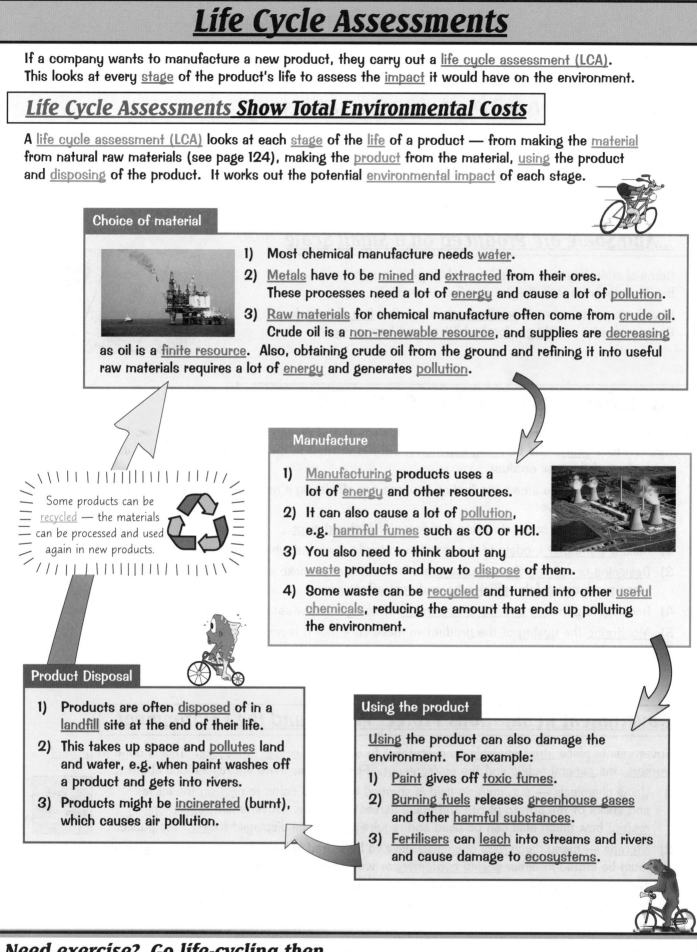

Choice of material

1) Most chemical manufacture needs water.

2) Metals have to be mined and extracted from their ores. These processes need a lot of energy and cause a lot of pollution.

3) Raw materials for chemical manufacture often come from crude oil. Crude oil is a non-renewable resource, and supplies are decreasing as oil is a finite resource. Also, obtaining crude oil from the ground and refining it into useful raw materials requires a lot of energy and generates pollution.

Some products can be recycled — the materials can be processed and used again in new products.

Manufacture

1) Manufacturing products uses a lot of energy and other resources.

2) It can also cause a lot of pollution, e.g. harmful fumes such as CO or HCl.

3) You also need to think about any waste products and how to dispose of them.

4) Some waste can be recycled and turned into other useful chemicals, reducing the amount that ends up polluting the environment.

Product Disposal

1) Products are often disposed of in a landfill site at the end of their life.

2) This takes up space and pollutes land and water, e.g. when paint washes off a product and gets into rivers.

3) Products might be incinerated (burnt), which causes air pollution.

Using the product

Using the product can also damage the environment. For example:

1) Paint gives off toxic fumes.

2) Burning fuels releases greenhouse gases and other harmful substances.

3) Fertilisers can leach into streams and rivers and cause damage to ecosystems.

Need exercise? Go life-cycling then...

Life cycle assessments are really good for evaluating the different materials you could use for a particular product. Once you know what needs to be considered at each stage you can evaluate the materials one by one. And you can also evaluate the product as a whole. Easy lemon squeezy peasy. Or something like that.

Section Twelve — Chemical Production

Synthesising Compounds

As I'm sure you know, the <u>chemical industry</u> is really <u>important</u>. Without it we'd be without loads of <u>everyday chemicals</u>. When it comes to <u>making</u> these chemicals it's not just a case of throwing everything into a bucket — oh no, there are quite a few <u>stages</u> to the <u>process</u> — <u>seven</u>, to be precise.

There are Seven Stages Involved in Chemical Synthesis

① CHOOSING THE REACTION

Chemists need to choose the reaction (or series of reactions) to make the product. For example:

- <u>neutralisation</u> (see p. 79) — an acid and an alkali react to produce a <u>salt</u>.
- <u>thermal decomposition</u> — heat is used to break up a compound into <u>simpler substances</u>.
- <u>precipitation</u> — an <u>insoluble solid</u> is formed when <u>two solutions</u> are mixed.

② RISK ASSESSMENT

This is an assessment of anything in the process that could <u>cause injury</u>. It involves:
- <u>identifying hazards</u>
- assessing <u>who might be harmed</u>
- deciding what <u>action</u> can be taken to <u>reduce the risk</u>.

③ CALCULATING THE QUANTITIES OF REACTANTS

This includes a lot of <u>maths</u> and a <u>balanced symbol equation</u>. Using the equation chemists can <u>calculate</u> how much of each <u>reactant</u> is needed to produce a <u>certain amount</u> of <u>product</u>. This is particularly important in industry because you need to know how much of each <u>raw material</u> is needed so there's no <u>waste</u> — waste costs <u>money</u>.

④ CHOOSING THE APPARATUS AND CONDITIONS

The reaction needs to be carried out using suitable <u>apparatus</u> and in the right <u>conditions</u>. The apparatus needs to be the <u>correct size</u> (for the <u>amount</u> of product and reactants) and <u>strength</u> (for the <u>type</u> of reaction being carried out, e.g. if it is explosive or gives out a lot of heat). Chemists need to decide what <u>temperature</u> the reaction should be carried out at, what <u>concentrations</u> of reactants should be used, and whether or not to use a <u>catalyst</u> (see p. 92-93).

⑤ ISOLATING THE PRODUCT

After the reaction is finished the products may need to be separated from the <u>reaction mixture</u>. This could involve <u>evaporation</u> (if the product is dissolved in the reaction mixture), <u>filtration</u> (if the product is an <u>insoluble solid</u>) and <u>drying</u> (to remove any <u>water</u>).

⑥ PURIFICATION

Isolating the product and <u>purification</u> go together like peas and carrots. As you're isolating the product you're also helping to purify it. <u>Crystallisation</u> can be useful in the purification process.

⑦ MEASURING YIELD AND PURITY

The yield tells you about the <u>overall success</u> of the process. It compares what you think you should get with what you get in practice (see p. 76). The purity of the chemical also needs to be measured (p. 109).

It's just like Snow White — but with chemical synthesis steps...

It's important that none of these stages are <u>missed out</u>. It'd be pointless if you weren't able to <u>separate</u> the product from the <u>reaction mixture</u> and even worse if the process caused <u>injury</u> or <u>death</u>. Be safe.

Producing Chemicals

Producing chemicals is a complicated business. Luckily most processes involve the same stages.

There Are Several Stages Involved in Producing Chemicals

The process of producing a useful chemical from the raw materials can be split into five stages:

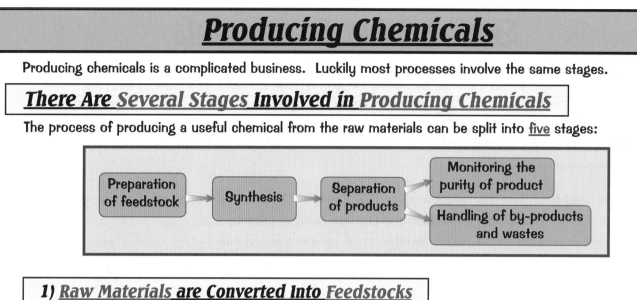

Preparation of feedstock → Synthesis → Separation of products → Monitoring the purity of product / Handling of by-products and wastes

1) Raw Materials are Converted Into Feedstocks

1) Raw materials are the naturally occurring substances which are needed, e.g. crude oil, natural gas.

2) Feedstocks are the actual reactants needed for the process, e.g. hydrogen, ethanol.

3) The raw materials usually have to be purified or changed in some way to make the feedstock.

2) Synthesis

The feedstocks (reactants) are converted by the magic of chemistry into products. The conditions have to be carefully controlled to make sure the reaction happens, and at a sensible rate.

3) The Products are Separated

1) Chemical reactions usually produce the substance you want and some other chemicals known as by-products. The by-products might be useful, or they might be waste.

2) You might also have some left-over reactants.

3) Everything has to be separated out so it can be dealt with in different ways.

4) The Purity of the Product is Monitored

1) Even after the best efforts are made to separate the product from everything else, it sometimes still has other things mixed in with it — it's not completely pure.

2) The purity of the product has to be monitored to make sure it's between certain levels.

3) Different industries need different levels of purity depending on what the product is used for. If a slightly impure product will do the job it's meant for, there's no point wasting money on purification.

5) By-products and Waste are Dealt With

1) Where possible, by-products are sold or used in another reaction.

2) If the reaction is exothermic, there may be waste heat. Heat exchangers can use excess heat to produce steam or hot water for other reactions — saving energy and money.

3) Waste products have to be carefully disposed of so they don't harm people or the environment — there are legal requirements about this.

Feedsock — when you spill your dinner over your feet...

As you might have noticed, producing chemicals isn't the most exciting of topics — but it is important. The chemical industry in Britain alone is worth billions of pounds... which suddenly makes it a lot more interesting.

Producing Chemicals

It'd be great if all industrial reactions were <u>sustainable</u> — humans could go <u>on and on</u> making whatever they wanted <u>forever</u> and ever. Life <u>isn't</u> like that though — so it's important to think about sustainability.

There Are Eight Key Questions About Sustainability

<u>Sustainable processes</u> are ones that <u>meet people's needs today without affecting the ability of future generations to meet their own needs</u>. Lots of factors affect whether a chemical process is sustainable.

① WILL THE RAW MATERIALS RUN OUT?

It's great if your feedstock is <u>renewable</u> — you can keep on using as much as you like. The trouble is, if it's not renewable it's going to <u>run out</u>. And this could mean <u>big problems</u> for future generations.

② HOW GOOD IS THE ATOM ECONOMY?

The <u>atom economy</u> of a reaction tells you how much of the <u>mass</u> of the reactants ends up as useful products. Pretty obviously, if you're making <u>lots of waste</u>, that's a <u>problem</u> — it all has to go somewhere. Reactions with low atom economy <u>use up resources</u> very quickly too.

③ WHAT DO I DO WITH THE WASTE PRODUCTS?

Waste products can be expensive to <u>remove</u> and dispose of <u>responsibly</u>. They're likely to <u>take up space</u> and cause <u>pollution</u>. One way around the problem is to find a <u>use</u> for the waste products rather than just <u>throwing them away</u>. Alternatively, there's often <u>more than one way</u> to make the product you want, so you try to choose a reaction that gives <u>useful by-products</u>.

④ WHAT ARE THE ENERGY COSTS?

If a reaction needs a lot of <u>energy</u> it'll be very <u>expensive</u>. And providing energy often involves <u>burning fossil fuels</u> — which of course is no good for the <u>environment</u>. But if a process <u>gives out</u> energy there might be a way to <u>use</u> that energy for something else — <u>saving money</u> and the <u>environment</u>.

⑤ WILL IT DAMAGE THE ENVIRONMENT?

Clearly if the reaction produces <u>harmful chemicals</u> it's <u>not</u> going to do any good for the <u>environment</u>. But you need to consider where the <u>raw materials</u> come from too (mining, for example, can make a right mess of the countryside, see p. 25), and also whether the reactants or products need <u>transporting</u>.

⑥ WHAT ARE THE HEALTH AND SAFETY RISKS?

There's no doubt about it — chemistry can be <u>dangerous</u>. There are <u>laws</u> in place that companies must follow to make sure their workers and the public are <u>not</u> put in harm's way. Companies also have to <u>test</u> their products to make sure they're <u>safe to use</u>.

⑦ ARE THERE ANY BENEFITS OR RISKS TO SOCIETY?

A factory creates <u>jobs</u> for the local community and brings <u>money</u> into the area. But it may be <u>unsightly</u> and potentially <u>hazardous</u>.

⑧ IS IT PROFITABLE?

This is the big question for most companies — businesses are out to make money after all. If the <u>costs</u> of a process are <u>higher</u> than the <u>income</u> from it, then it won't be <u>profitable</u>.

Enough with the questions — I confess...

Hmm, if you want to make a product safely and sustainably than you have to answer these eight questions. If this is still all a bit confusing there are some <u>examples</u> of how to go about it on pages 127, 129 and 130.

The Haber Process

This is an <u>important industrial process</u>. It produces <u>ammonia</u> (NH_3), which is used to make <u>fertilisers</u>.

Nitrogen and Hydrogen Are Needed to Make Ammonia

$$N_{2 (g)} + 3H_{2 (g)} \rightleftharpoons 2NH_{3 (g)} \quad (+ \text{ heat})$$

1) The <u>nitrogen</u> is obtained easily from the <u>air</u>, which is <u>78% nitrogen</u> (and 21% oxygen). *These gases are first purified.*

2) The <u>hydrogen</u> comes from <u>natural gas</u> or from <u>other sources</u> like crude oil.

3) Some of the nitrogen and hydrogen reacts to form <u>ammonia</u>. Because the reaction is <u>reversible</u> — it occurs in both directions — ammonia breaks down again into nitrogen and hydrogen. The reaction reaches an <u>equilibrium</u>.

> **Industrial conditions:**
> <u>Pressure</u>: 200 atmospheres (atm);   <u>Temperature</u>: 450 °C;   <u>Catalyst</u>: Iron

The Reaction is Reversible, So There's a Compromise to be Made:

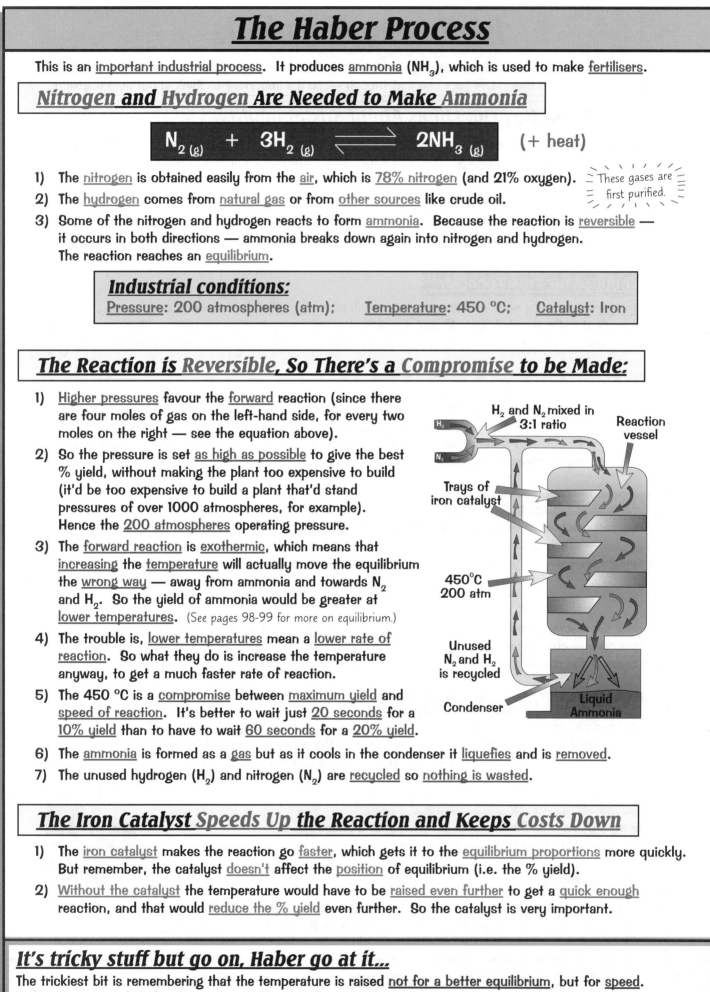

1) <u>Higher pressures</u> favour the <u>forward</u> reaction (since there are four moles of gas on the left-hand side, for every two moles on the right — see the equation above).

2) So the pressure is set <u>as high as possible</u> to give the best % yield, without making the plant too expensive to build (it'd be too expensive to build a plant that'd stand pressures of over 1000 atmospheres, for example). Hence the <u>200 atmospheres</u> operating pressure.

3) The <u>forward reaction</u> is <u>exothermic</u>, which means that <u>increasing</u> the <u>temperature</u> will actually move the equilibrium the <u>wrong way</u> — away from ammonia and towards N_2 and H_2. So the yield of ammonia would be greater at <u>lower temperatures</u>. (See pages 98-99 for more on equilibrium.)

4) The trouble is, <u>lower temperatures</u> mean a <u>lower rate of reaction</u>. So what they do is increase the temperature anyway, to get a much faster rate of reaction.

5) The 450 °C is a <u>compromise</u> between <u>maximum yield</u> and <u>speed of reaction</u>. It's better to wait just <u>20 seconds</u> for a <u>10% yield</u> than to have to wait <u>60 seconds</u> for a <u>20% yield</u>.

Labels in diagram: H_2 and N_2 mixed in 3:1 ratio; Reaction vessel; Trays of iron catalyst; 450°C 200 atm; Unused N_2 and H_2 is recycled; Condenser; Liquid Ammonia

6) The <u>ammonia</u> is formed as a <u>gas</u> but as it cools in the condenser it <u>liquefies</u> and is <u>removed</u>.

7) The unused hydrogen (H_2) and nitrogen (N_2) are <u>recycled</u> so <u>nothing is wasted</u>.

The Iron Catalyst Speeds Up the Reaction and Keeps Costs Down

1) The <u>iron catalyst</u> makes the reaction go <u>faster</u>, which gets it to the <u>equilibrium proportions</u> more quickly. But remember, the catalyst <u>doesn't</u> affect the <u>position</u> of equilibrium (i.e. the % yield).

2) <u>Without the catalyst</u> the temperature would have to be <u>raised even further</u> to get a <u>quick enough</u> reaction, and that would <u>reduce the % yield</u> even further. So the catalyst is very important.

It's tricky stuff but go on, Haber go at it...

The trickiest bit is remembering that the temperature is raised <u>not for a better equilibrium</u>, but for <u>speed</u>. It doesn't matter that the percentage yield is low, because the hydrogen and nitrogen are recycled. Cover the page and scribble down as much as you can remember, then check, and try again.

Nitrogen Fixation

Nitrogen fixation is about taking nitrogen and turning it into chemicals we can use. Just like in the Haber process...

Nitrogen Fixation Turns N₂ from the Air into Ammonia

1) Nitrogen fixation is the process of turning N_2 from the air into useful nitrogen compounds like ammonia.

2) The Haber process is a non-biological way of fixing nitrogen (see previous page).

3) Most of the ammonia produced by the Haber process is used to make fertilisers.

4) Fertilisers play a vital part in world food production as they increase crop yield so help to feed more people.

5) When used in large amounts though, fertilisers can pollute water supplies and cause eutrophication (see page 118).

6) Ammonia is also very important in industry where it is used to manufacture plastics, explosives and pharmaceuticals.

The Efficiency of Nitrogen Fixation can be Improved by Catalysts

1) In the Haber process very high temperatures and pressures have to be used to turn nitrogen and hydrogen into ammonia.

2) Using an iron catalyst makes the rate of reaction much faster — so the ammonia is produced faster.

3) Without the catalyst the temperature would have to be raised even further to get a quick enough reaction, and that would reduce the % yield even further. So the catalyst is very important.

4) Some living organisms such as nitrogen-fixing bacteria can fix nitrogen at room temperature and pressure. They do this using biological catalysts called enzymes.

5) Chemists would like to be able to make catalysts that mimic these enzymes, so that processes like the Haber process can be carried out at room temperature and pressure.

6) As it's expensive and time consuming to work at high temperatures and pressures, this would mean that processes involving nitrogen fixation would become much cheaper and more efficient. Hooray.

Is Nitrogen Fixation A Sustainable Process?

Here's a run down of the sustainability of the Haber process.

1) Will the raw materials run out? — Hydrogen comes from fossil fuels. They're non-renewable and will run out. Nitrogen comes from the air so it's very unlikely it will run out.

2) How good is the atom economy? — All the H_2 and N_2 makes ammonia so the atom economy is excellent.

3) What do I do with my waste products? — There are no waste products as the chemicals are all recycled.

4) What are the energy costs? — Lots of energy is needed to keep the reaction at 450 °C and 200 atm.

5) Will it damage the environment? — Fertilisers made from NH_3 can cause eutrophication and water pollution.

6) What are the health and safety risks? — Working at high temperatures and pressures can be very dangerous.

7) Are there any benefits or risks to society? — Making ammonia can help world food production.

8) Is it profitable? — Yes, making ammonia is big business.

Daisy, Mary, Buttercup... — It's a cattle-list...

Mother Nature 1 - Humans 0. Here's yet another reason why us humans aren't quite as smart as we'd like to think. If I could invent a process which was half as efficient as those nitrogen-fixing bacteria I'd have more money than I could shake a stick at. Instead I'll just have to revise hard, pass my exams and get rich that way.

The Contact Process

And here's another example where getting the conditions right makes you <u>more product</u>. Whoop.

The Contact Process <u>is Used to Make</u> Sulfuric Acid

1) The first stage is to make <u>sulfur dioxide</u> (SO_2) — usually by burning <u>sulfur</u> in <u>air</u>.

> sulfur + oxygen → sulfur dioxide
> $S_{(s)} + O_{2(g)} \rightarrow SO_{2(g)}$

This isn't the contact process I wanted to learn about...

2) The sulfur dioxide is then <u>oxidised</u> (with the help of a catalyst) to make <u>sulfur trioxide</u> (SO_3).

> sulfur dioxide + oxygen $\rightleftharpoons$ sulfur trioxide
> $2SO_{2(g)} + O_{2(g)} \rightleftharpoons 2SO_{3(g)}$

3) Next, the sulfur trioxide is used to make <u>sulfuric acid</u>.

> sulfur trioxide + water → sulfuric acid
> $SO_{3(g)} + H_2O_{(l)} \rightarrow H_2SO_{4(aq)}$

In reality, dissolving SO_3 like this doesn't work — the reaction is dangerous as a lot of heat's produced. (In practice, you dissolve SO_3 in sulfuric acid first.)

The <u>Conditions</u> <u>Used to Make</u> SO$_3$ are <u>Carefully Chosen</u>

The reaction in step 2 is <u>reversible</u>. So, the <u>conditions</u> used can be <u>controlled</u> to get a <u>higher yield</u> (more product).

> $2SO_2 + O_2 \rightleftharpoons 2SO_3$

The forward reaction is exothermic.

TEMPERATURE

1) Oxidising sulfur dioxide to form sulfur trioxide is <u>exothermic</u> (it <u>gives out</u> heat).

2) So to get <u>more product</u> you'd think the temperature should be <u>reduced</u> (so the equilibrium will shift to the <u>right</u> to <u>replace the heat</u>).

3) Unfortunately, reducing the temperature <u>slows</u> the reaction right down — not much good.

4) So a <u>compromise</u> temperature of <u>450 °C</u> is used — to get quite a high yield quite quickly.

PRESSURE

1) There are <u>two moles</u> of <u>product</u>, compared to <u>three moles</u> of <u>reactants</u>.

2) So to get <u>more product</u>, you'd think the pressure should be <u>increased</u> (so that the equilibrium will shift to the <u>right</u> to <u>reduce the pressure</u>).

3) But increasing the pressure is <u>expensive</u>, and as the equilibrium is already on the right, it's not really <u>necessary</u>.

4) In fact, <u>atmospheric pressure</u> (1 atmosphere) is used.

CATALYST

1) To <u>increase</u> the rate of reaction a <u>vanadium pentoxide catalyst</u> (V_2O_5) is used.

2) It <u>DOESN'T</u> change the <u>position</u> of the equilibrium.

With a <u>fairly high temperature</u>, a <u>low pressure</u> and a <u>vanadium pentoxide catalyst</u>, the reaction goes <u>pretty quickly</u> and you get a <u>good yield</u> of SO_3 (about 99%).

And that's how chemistry works in real life.

The lonely hearts column — go on, start the contact process...

It's a tough one... do you <u>raise</u> the <u>temperature</u> to get a <u>faster</u> rate of reaction, or <u>reduce</u> it to get a better <u>yield</u>... In the end you <u>compromise</u> (as is so often the case in life... sigh). And that's before you even start to worry about the <u>cost</u> of raising the temperature. Decisions, decisions...

Section Twelve — Chemical Production

Making Ethanol

Ethanol is the alcohol people <u>drink</u> — but as you saw on page 41 this is far from its only use. It's also used as a <u>fuel</u>, a <u>solvent</u> and as a <u>feedstock</u> for other processes.

Ethanol can be Made by Fermentation

The ethanol in <u>alcoholic drinks</u> is usually made using fermentation.

1) <u>Fermentation</u> uses <u>yeast</u> to convert <u>sugars</u> into <u>ethanol</u>. Carbon dioxide is also produced.

$$\text{sugar} \xrightarrow{\text{yeast}} \text{ethanol + carbon dioxide}$$

2) The yeast cells contain <u>zymase</u>, an <u>enzyme</u> that acts as a catalyst in fermentation.

3) Fermentation happens fastest at about <u>30 °C</u>. That's because zymase works best at this temperature. At lower temperatures, the reaction slows down. And if it's <u>too hot</u> the zymase is <u>denatured</u> (destroyed).

4) Zymase also works best at a <u>pH</u> of about <u>4</u> — a strongly acidic or alkaline solution will stop it working.

5) It's important to prevent <u>oxygen</u> getting to the fermentation process. Oxygen converts the <u>ethanol</u> to <u>ethanoic acid</u> (the acid in <u>vinegar</u>), which lowers the pH and can stop the enzyme working.

6) When the <u>concentration</u> of ethanol reaches about 10 to 20%, the fermentation reaction <u>stops</u>, because the yeast gets <u>killed off</u> by the ethanol.

Ethanol Solution can be Concentrated by Distillation

The fermented mixture can be <u>distilled</u> to produce more <u>concentrated</u> ethanol, which can then be used to make products like <u>brandy</u> or <u>whisky</u>.

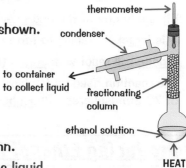

1) The ethanol solution is put in a flask below a <u>fractionating column</u>, as shown.

2) The solution is <u>heated</u> so that the <u>ethanol</u> boils. The ethanol vapour travels <u>up</u> the column, <u>cooling</u> down as it goes.

3) The temperature is such that anything with a <u>higher</u> boiling point than ethanol (like water) cools to a <u>liquid</u> and flows <u>back</u> into the <u>solution</u> at the bottom.

4) This means that <u>only</u> pure ethanol vapour reaches the <u>top</u> of the column.

5) The ethanol vapour flows through a <u>condenser</u> — where it's cooled to a <u>liquid</u>, which is then collected.

Is Fermentation A Sustainable Process?

You can work out if fermentation is <u>sustainable or not</u> by answering the <u>eight questions</u> from page 125.

1) <u>Will the raw materials run out?</u> — Sugar beet and yeast <u>grow quickly</u> so <u>won't</u> run out.

2) <u>How good is the atom economy?</u> — The <u>waste CO_2</u> produced means it has a <u>low atom economy</u>. And because the enzyme is <u>killed off</u> by the ethanol produced, the reaction is even <u>less</u> efficient.

3) <u>What do I do with my waste products?</u> — The <u>waste CO_2</u> can be <u>released</u> without any processing.

4) <u>What are the energy costs?</u> — <u>Energy</u> is needed to keep the reaction at its <u>optimum</u> temperature.

5) <u>Will it damage the environment?</u> — <u>Carbon dioxide</u> is a <u>greenhouse gas</u> so adds to <u>global warming</u>.

6) <u>What are the health and safety risks?</u> — The chemicals and processes do <u>not</u> have any specific dangers.

7) <u>Are there any benefits or risks to society?</u> — Making ethanol <u>doesn't</u> impact society (drinking it does).

8) <u>Is it profitable?</u> — This depends on what the ethanol is <u>used</u> for, e.g. <u>drinking</u> or <u>fuel</u>.

Excessive drinking — when a tipple becomes a topple...

People have been making alcohol for thousands of years. It could explain why there are so many ancient ruins all over the place — maybe the Romans were always <u>too drunk</u> to finish the job properly...

Making Ethanol

Ethanol can be Made From Biomass

Scientists have recently developed a way to make ethanol from waste biomass.

1) Waste biomass is the parts of a plant that would normally be thrown away — e.g. corn stalks, rice husks, wood pulp and straw.

2) Waste biomass cannot be fermented in the normal way because it contains a lot of cellulose. Yeast can easily convert some sugars to ethanol, but it can't convert cellulose to ethanol.

3) E. coli bacteria can be genetically modified so they can convert cellulose in waste biomass into ethanol.

4) The optimum conditions for this process are a temperature of 35 °C and a slightly acidic solution, pH 6.

Is Producing Ethanol from Biomass a Sustainable Process?

The sustainability of the biomass method is very similar to the sustainability of the standard fermentation method because they both use similar processes. The advantage of using biomass is that you don't have to grow crops specially for producing ethanol — you can use the waste from other crops.

Ethene Can be Reacted with Steam to Produce Ethanol

Fermentation is too slow for making ethanol on a large scale. Instead, ethanol is made on an industrial scale using ethane. This method allows high quality ethanol to be produced continuously and quickly.

1) Ethane is one of the hydrocarbons found in crude oil and natural gas.

2) It is 'cracked' (split) to form ethene (C_2H_4) and hydrogen gas.

3) Ethene will react with steam (H_2O) to make ethanol.

ethane → ethene + hydrogen

ethene + steam → ethanol

4) The reaction needs a temperature of 300 °C and a pressure of 70 atmospheres. Phosphoric acid is used as a catalyst.

Is Producing Ethanol from Ethane a Sustainable Process?

You know the drill by now — answer the eight questions from page 125 to work out if making ethanol from ethane is sustainable or not. You can only answer them if you know enough about the process though.

1) Will the raw materials run out? — Crude oil and natural gas are non-renewable and will run out.

2) How good is the atom economy? — Cracking ethane has a fairly high atom economy as the only waste product is hydrogen. Reacting ethene has an even higher atom economy as ethanol is the only product.

3) What do I do with my waste products? — The only waste is the hydrogen gas produced by cracking ethane. It can be reused to make ammonia in the Haber process.

4) What are the energy costs? — Energy is needed to maintain the high temperature and pressure used.

5) Will it damage the environment? — The reactions involved do not produce any waste products that directly harm the environment. But, crude oil can harm the environment, e.g. through oil spills.

6) What are the health and safety risks? — The high temperature and pressure used to produce the ethanol have to be carefully controlled — otherwise it could be very dangerous.

7) Are there any benefits or risks to society? — This method has no specific impact on society.

8) Is it profitable? — Yes, manufacturing ethanol from ethene and steam is continuous and quick and the raw materials are fairly cheap — but it won't stay that way once crude oil starts to run out.

Biomass — an organic church...

It's important to think about the sustainability of chemical processes — it'd be rather selfish to use up all the raw materials, and leave future generations with big piles of our waste to put up with instead.

Making Esters

You'll probably remember <u>esters</u> from page 43... well, here's <u>a bit more</u> about them.

How to Make an Ester — <u>Reflux, Distil, Purify, Dry</u>

Making esters is a bit more complicated than just mixing an alcohol and a carboxylic acid together. The reaction is <u>reversible</u> so some of the <u>ester</u> will react with the other product (<u>water</u>), and <u>re-form</u> the <u>carboxylic acid</u> and <u>alcohol</u>. To get a <u>pure</u> ester you need a multi-step <u>reaction</u> and <u>purification</u> procedure.

1) <u>Refluxing</u> — <u>The Reaction</u>

To make ethyl ethanoate you need to react ethanol with ethanoic acid, using a <u>catalyst</u> such as <u>concentrated sulfuric acid</u> to speed things up.

<u>Heating</u> the mixture also speeds up the reaction — but you can't just stick a Bunsen under it as the ethanol will evaporate or catch fire before it can react.

Instead, the mixture's <u>gently heated in a flask</u> fitted with a <u>condenser</u> — this catches the vapours and <u>recycles</u> them back into the flask, giving them <u>time to react</u>. This handy method is called <u>refluxing</u>.

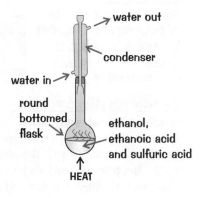

2) <u>Distillation</u>

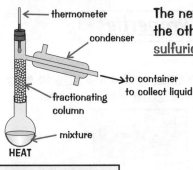

The next step is <u>distillation</u>. This <u>separates</u> your lovely <u>ester</u> from all the other stuff left in the flask (<u>unreacted alcohol</u> and <u>carboxylic acid</u>, <u>sulfuric acid</u> and <u>water</u>).

The mixture's <u>heated</u> below a fractionating column. As it starts to boil, the vapour goes up the <u>fractionating column</u>.

When the temperature at the top of the column reaches the <u>boiling point</u> of <u>ethyl ethanoate</u>, the liquid that flows out of the condenser is <u>collected</u>. This liquid is <u>impure</u> ethyl ethanoate.

3) <u>Purification</u>

The liquid collected (the <u>distillate</u>) is poured into a <u>tap funnel</u> and then treated to remove its impurities, as follows:

The mixture is shaken with <u>sodium carbonate solution</u> to remove acidic impurities. Ethyl ethanoate doesn't mix with the water in the sodium carbonate solution, so the mixture <u>separates</u> into two layers, and the lower layer can be <u>tapped off</u> (removed).

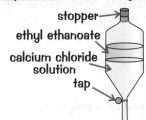

The remaining upper layer is then shaken with <u>concentrated calcium chloride</u> solution to remove any <u>ethanol</u>. Again, the lower layer can be <u>tapped off</u> and removed.

4) <u>Drying</u>

Any remaining <u>water</u> in the ethyl ethanoate can be removed by shaking it with lumps of <u>anhydrous calcium chloride</u>, which absorb the water — this is called <u>drying</u>. Finally, the <u>pure ethyl ethanoate</u> can be separated from the solid calcium chloride by <u>filtration</u>.

Esterification — it's 'esterical stuff...

Sorry about that, it's a little bit tricky really. The secret is to get your head around <u>each step</u> of the process before you try and <u>put it all together</u>. Read each step, cover, scribble... you know the drill.

Making Polymers

There's plastic and there's... well, plastic. You wouldn't want to make a chair with the same plastic that gets used for flimsy old carrier bags. But whatever the plastic, it's always a polymer.

Forces Between Molecules Determine the Properties of Plastics

Strong covalent bonds hold the atoms together in long chains. But it's the bonds between the different molecule chains that determine the properties of the plastic.

Weak Forces:
Individual tangled chains of polymers, held together by weak intermolecular forces, are free to slide over each other.

THERMOSOFTENING POLYMERS don't have cross-linking between chains. The forces between the chains are really easy to overcome, so it's dead easy to melt the plastic. When it cools, the polymer hardens into a new shape. You can melt these plastics and remould them as many times as you like.

Strong Forces:
Some plastics have stronger intermolecular forces between the polymer chains, called crosslinks, that hold the chains firmly together.

THERMOSETTING POLYMERS have crosslinks. These hold the chains together in a solid structure. The polymer doesn't soften when it's heated. Thermosetting polymers are the tough guys of the plastic world. They're strong, hard and rigid.

Polymers Can be Modified to Give Them Different Properties

You can chemically modify polymers to change their properties.

1) Polymers can be modified to increase their chain length. Polymers with short chains are easy to shape and have lower melting points. Longer chain polymers are stiffer and have higher melting points.

2) Polymers can be made stronger by adding cross-linking agents. These agents chemically bond the chains together, making the polymer stiffer, stronger and more heat-resistant.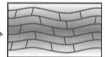

3) Plasticisers can be added to a polymer to make it softer and easier to shape. Plasticisers work by getting in between the polymer chains and reducing the forces between them.

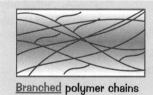 Branched polymer chains

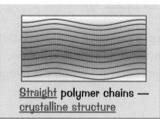 Straight polymer chains — crystalline structure

4) The polymer can be made more crystalline. A crystalline polymer has straight chains with no branches so the chains can fit close together. Crystalline polymers have higher density, are stronger and have a higher melting point.

How You Make a Polymer Affects Its Properties

1) The starting materials and reaction conditions will both affect the properties of a polymer.

2) Two types of polythene can be made using different conditions:

- Low density (LD) polythene is made by heating ethene to about 200 °C under high pressure. It's flexible and is used for bags and bottles.

- High density (HD) polythene is made at a lower temperature and pressure (with a catalyst). It's more rigid and is used for water tanks and drainpipes.

Choose your polymers wisely...

The molecules that make up a plastic affect the properties of the plastic, which also affects what the plastic can be used for. If they're not useful enough, you can modify the polymers to make them even more useful. Hurrah.

New Materials

New materials are continually being developed, with new properties. The two groups of materials you really need to know about are <u>smart materials</u> and <u>nanoparticles</u>.

Smart Materials Have Some Really Weird Properties

1) <u>Smart</u> materials <u>behave differently</u> depending on the <u>conditions</u>, e.g. temperature.

2) A good example is <u>nitinol</u> — a "<u>shape memory alloy</u>".
It's a metal <u>alloy</u> (about half nickel, half titanium) but when it's cool you can <u>bend it</u> and <u>twist it</u> like rubber. Bend it too far, though, and it stays bent. But here's the really clever bit — if you heat it above a certain temperature, it goes back to a "<u>remembered</u>" shape.

3) It's really handy for <u>glasses frames</u>. If you accidentally bend them, you can just pop them into a bowl of hot water and they'll <u>jump</u> back <u>into shape</u>.

4) Nitinol is also used for <u>dental braces</u>. In the mouth it <u>warms</u> and tries to return to a 'remembered' shape, and so it gently <u>pulls the teeth</u> with it.

Nanoparticles Are Really Really Really Really Tiny   ...smaller than that.

1) Really tiny particles, <u>1–100 nanometres</u> across, are called 'nanoparticles' (1 nm = 0.000 000 001 m).

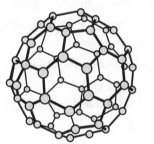

2) Nanoparticles contain roughly <u>a few hundred atoms</u>.

3) Nanoparticles include <u>fullerenes</u>. These are molecules of <u>carbon</u>, shaped like <u>hollow balls</u> or <u>closed tubes</u>. The carbon atoms are arranged in <u>hexagonal rings</u>. Different fullerenes contain <u>different numbers</u> of carbon atoms.

4) A nanoparticle has very <u>different properties</u> from the 'bulk' chemical that it's made from — e.g. <u>fullerenes</u> have different properties from big <u>lumps of carbon</u>.

Fullerenes Can Form Very Strong Nanotubes

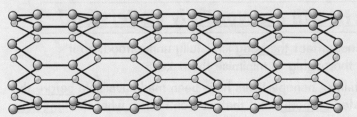

1) Fullerenes can be joined together to form <u>nanotubes</u> — teeny tiny hollow carbon tubes, a few nanometres across.

2) All those covalent bonds make carbon nanotubes <u>very strong</u>.

Bendy specs, tennis rackets and computer chips — cool...

The phrase "good things come in little packages" was made for <u>nanoparticles</u> (and Tiger — my poor hamster who is no more. It's a shame his dash for freedom took him under the wheels of my sister's bike. Sad stuff.)

Section Twelve — Chemical Production

Nanotechnology

This page is all about the <u>uses</u> of those handy nanoparticles from the last page.

Nanomaterials *are* Often *Designed* for a *Specific Use*

Most nanomaterials are <u>made</u> using nanotechnology, but some nanoscale materials <u>occur naturally</u> or are produced by <u>accident</u>. For example:

- <u>Seaspray</u> — The sea produces nanoscale <u>salt particles</u> which are present in the <u>atmosphere</u>.
- <u>Combustion</u> — When <u>fuels</u> are burnt, nanoscale <u>soot particles</u> are produced.

Nanoparticles are Often Used to *Modify* the Properties of *Materials*

Using nanoparticles is known as <u>nanoscience</u>. Many <u>new uses</u> of nanoparticles are being developed:

- They have a <u>huge surface area to volume ratio</u>, so they could help make new industrial <u>catalysts</u> (see page 93).
- You can use nanoparticles to make <u>sensors</u> to detect one type of molecule and nothing else. These <u>highly specific</u> sensors are already being used to test water purity.
- Nanotubes can be used to make <u>stronger</u>, <u>lighter</u> building materials.
- New cosmetics, e.g. <u>sun tan cream</u> and <u>deodorant</u>, have been made using nanoparticles. The small particles do their job but don't leave <u>white marks</u> on the skin.
- <u>Nanomedicine</u> is a hot topic. The idea is that tiny fullerenes are <u>absorbed</u> more easily by the body than most particles. This means they could <u>deliver drugs</u> right into the cells where they're needed.
- New <u>lubricant coatings</u> are being developed using fullerenes. These coatings reduce friction a bit like <u>ball bearings</u> and could be used in all sorts of places from <u>artificial joints</u> to <u>gears</u>.
- Nanotubes <u>conduct</u> electricity, so they can be used in tiny <u>electric circuits</u> for computer chips.
- Nanoparticles are added to <u>plastics</u> in <u>sports equipment</u>, e.g. tennis rackets, golf clubs and golf balls. They make the plastic much <u>stronger</u> and <u>more durable</u>, and they don't <u>add weight</u>.
- <u>Silver nanoparticles</u> are added to <u>polymer fibres</u> used to make <u>surgical masks</u> and <u>wound dressings</u>. This gives the fibres <u>antibacterial properties</u>.

Nanoparticles are much <u>smaller</u> than larger particles of the <u>same material</u>. This means they have a <u>larger surface area-to-volume ratio</u>, which is what gives them <u>different properties</u> and makes them super <u>useful</u>. <u>Silver nanoparticles</u>, for example, can <u>kill bacteria</u>, making them suitable for wound dressings. <u>Normal silver particles</u> are much bigger, have a smaller surface area to volume ratio and <u>can't</u> kill bacteria. Not so useful. Well, not for wound dressings anyway.

The Effects of Nanoparticles on *Health* *are Not Fully* Understood

1) Although nanoparticles are useful, the way they affect <u>the body</u> isn't fully understood, so it's important that any new products are <u>tested</u> thoroughly to minimise the risks.

2) Some people are worried that <u>products</u> containing nanoparticles have been made available <u>before</u> the effects on <u>human health</u> have been investigated <u>properly</u>, and that we don't know what the <u>long-term</u> impacts on health will be.

3) As the long-term impacts aren't known, many people believe that products containing nanoscale particles should be <u>clearly labelled</u>, so that consumers can choose whether or not to use them.

Carbon nanoparticles — bit of a fancy name for soot...

Some nanoparticles have really <u>unexpected properties</u>. Gold nanoparticles aren't gold-coloured, for example — they're either red or purple. Now that's definitely unexpected. But not all that <u>useful</u>. But others, such as antibacterial silver nanoparticles, are both unexpected <u>and</u> useful. Hurrah for unexpectedly useful properties.

CFCs and the Ozone Layer

Scientists <u>changed their minds</u> about CFCs as they found out more evidence about them.

At First **Scientists Thought** CFCs were Great...

1) <u>Chlorofluorocarbons</u> (CFCs for short) are <u>organic molecules</u> containing <u>carbon</u>, <u>chlorine</u> and <u>fluorine</u>, e.g. <u>dichlorodifluoromethane</u> CCl_2F_2 — this is like <u>methane</u> but with two chlorine and two fluorine atoms (and an extremely long name) instead of the four hydrogen atoms.

2) CFCs are <u>non-toxic</u>, <u>non-flammable</u> and chemically <u>inert</u> (unreactive). They're <u>insoluble</u> in water and have <u>low boiling points</u>. Scientists were very happy that they'd found some <u>non-toxic</u> and <u>inert</u> chemicals which were <u>ideal for many uses</u>.

3) Chlorofluorocarbons were used as <u>coolants</u> in <u>refrigerators</u> and <u>air-conditioning systems</u>.

4) CFCs were also used as <u>propellants</u> in <u>aerosol spray cans</u>.

...But Then They Discovered the Shocking Truth

This is called the "hole in the ozone layer".

1) In 1974 scientists found that <u>chlorine</u> could help to <u>destroy ozone</u> (see equations on p. 136).

2) In 1985 scientists found <u>evidence</u> of <u>decreasing ozone levels</u> in the atmosphere over Antarctica.

3) Measurements in the upper atmosphere show high levels of compounds produced by the <u>breakdown of CFCs</u>. This supports the hypothesis that CFCs break down and destroy ozone.

4) Scientists are now <u>sure</u> that CFCs are linked to the depletion (thinning) of the ozone layer.

- <u>Ozone</u> is a form of oxygen with the formula O_3 — it has <u>three oxygen atoms</u> per molecule, unlike ordinary oxygen which has <u>two</u> atoms per molecule.

- It hangs about in the <u>ozone layer</u>, way up in the <u>stratosphere</u> (part of the upper atmosphere), doing the very important job of <u>absorbing ultraviolet (UV) light</u> from the Sun. Ozone absorbs UV light and breaks down into an <u>oxygen molecule</u> and an <u>oxygen atom</u>: O_3 + UV light $\rightarrow$ $O + O_2$ The oxygen molecule and oxygen atom join together to <u>make ozone again</u>: $O + O_2 \rightarrow O_3$

- <u>Reducing</u> the amount of <u>ozone</u> in the stratosphere results in <u>more UV light</u> passing through the atmosphere. Increased levels of <u>UV light</u> hitting the surface of the Earth can cause <u>medical problems</u> like increased risk of <u>sunburn</u> and <u>skin cancer</u>.

Some Countries **Have Banned the Use of** CFCs

1) Scientists' view that CFCs could damage the ozone layer caused a lot of <u>concern</u>.

2) But it took a while for society to do something about the mounting scientific evidence. Governments waited until the research had been thoroughly <u>peer reviewed</u> and <u>evaluated</u> before making a decision.

3) In 1978 the USA, Canada, Sweden and Norway <u>banned CFCs as aerosol propellants</u>.

4) After the <u>ozone hole</u> was discovered many countries (including the UK) got together and decided to reduce CFC production and eventually <u>ban CFCs completely</u>.

That's "chlorofluorocarbon" not "Chelsea Football Club"...

The ozone in the stratosphere is amazing stuff — it absorbs UV light and stops us from having to bear the full force of the Sun's UV output. Too much UV causes <u>sunburn</u> and <u>skin cancer</u>, so anything that damages the ozone and lets more UV through is a <u>bad</u> thing in the long run. So no more CFCs in fridges and spray cans.

CFCs and the Ozone Layer

CFCs damage ozone by forming <u>free radicals</u>. Free radicals are made when <u>covalent bonds</u> split <u>evenly</u>...

Free Radicals <u>are Made by</u> Breaking Covalent Bonds

1) A <u>covalent bond</u>, remember, is one where <u>two atoms share electrons</u> between them, like in H_2 (page 64).

2) A covalent bond can <u>break unevenly</u> to form <u>two ions</u>, e.g. $H-H \rightarrow H^+ + H^-$.
The H^- has <u>both</u> of the shared electrons, and the poor old H^+ has <u>neither</u> of them.

3) But a covalent bond can also break <u>evenly</u> — and then <u>each atom</u> gets <u>one</u> of the shared electrons, e.g. $H-H \rightarrow H\cdot + H\cdot$ — the $H\cdot$ is called a <u>free radical</u>. (The unpaired electron is shown by a <u>dot</u>.)

4) The unpaired electron makes the free radical <u>very, very reactive</u>.

Chlorine Free Radicals <u>from</u> CFCs <u>Damage the</u> Ozone Layer

1) <u>Ultraviolet light</u> makes the carbon-chlorine bonds in CFCs break to form <u>free radicals</u>:

2) This happens <u>high up in the atmosphere</u> (in the <u>stratosphere</u>), where the <u>ultraviolet light</u> from the Sun is <u>stronger</u>.

$$CCl_2F_2 \rightarrow CClF_2\cdot + Cl\cdot$$

3) <u>Chlorine free radicals</u> from this reaction react with <u>ozone</u> (O_3), turning it into ordinary oxygen molecules (O_2) and chlorine oxide ($ClO\cdot$):

$$O_3 + Cl\cdot \rightarrow ClO\cdot + O_2$$

4) The chlorine oxide molecule is <u>very reactive</u>, and reacts with ozone to make two <u>oxygen molecules</u> and <u>another Cl· free radical</u>:

$$ClO\cdot + O_3 \rightarrow 2O_2 + Cl\cdot$$

5) This Cl· free radical now goes and reacts with <u>another ozone molecule</u>. This is a <u>chain reaction</u>, so just <u>one chlorine free radical</u> from one CFC molecule can go around breaking up <u>a lot of ozone molecules</u>.

> CFCs <u>don't attack ozone directly</u>. They break up and form chlorine atoms (chlorine free radicals) which attack ozone. The chlorine atoms <u>aren't used up</u>, so they can carry on breaking down ozone.

CFCs <u>Stay in the Stratosphere for</u> Ages

1) CFCs are <u>not very reactive</u> and will only react with one or two chemicals that are present in the atmosphere. And they'll only break up to form <u>chlorine atoms</u> in the stratosphere, where there's plenty of high-energy ultraviolet light around. They won't do it in the lower atmosphere.

2) This means that the CFCs in the stratosphere now will take a <u>long time</u> to be removed.

3) Remember, each CFC molecule produces one chlorine atom which can react with an <u>awful lot</u> of ozone molecules. <u>Thousands</u> of them, in fact.

4) So the millions of CFC molecules that are present in the stratosphere will continue to destroy ozone for a long time — even <u>after all CFCs have been banned</u>. Each molecule will <u>stay around</u> for a long time, and each molecule will <u>destroy a lot of ozone</u> molecules.

Alkanes <u>and</u> HFCs <u>are</u> Safe Alternatives <u>to</u> CFCs

1) Alkanes <u>don't react</u> with ozone, so they can provide a safe alternative to CFCs.

2) <u>Hydrofluorocarbons</u> (<u>HFC</u>s) are compounds very similar to CFCs — but they contain <u>no chlorine</u>. It's the chlorine in CFCs that attacks ozone, remember.

3) Scientists have investigated the compounds that could be produced by breakdown of HFCs in the upper atmosphere, and <u>none of them</u> seem to be able to <u>attack ozone</u>. <u>Evidence suggests</u> HFCs are <u>safe</u>.

Oooh, here comes the tricky science bit...

I saved the worst till last I'm afraid with this page. You can't just glide your eyes over the equations and hope for the best. <u>Cover the page and scribble them down</u>, then check what you wrote. It'll be worth it in the end.

Revision Summary for Section Twelve

The end of another beautiful section and the end of the whole book — it brings a tear to my eye. Here's a handy pocket-sized checklist of things to make sure you've learnt: 1. definitions — learn what all the words mean, yes even the really long ones, 2. formulas — you'll lose easy marks on calculations if you don't, 3. examples — examiners love giving you marks for dropping the odd example in here and there, 4. reactions — these are the bread and butter of chemistry, 5. how to do things — diagrams often help here...
I can't think of anything else right now, so try these questions to check I haven't forgotten anything vital.

1) What is meant by the term 'bulk chemical'? Give two examples of bulk chemicals.
2) What are 'fine chemicals'? Give an example.
3) What does LCA stand for?
4) What are the four stages that need to be considered when doing an LCA?
5) When designing a chemical process what is involved in carrying out a risk assessment?
6) Why is it important to choose the right apparatus for a chemical process?
7) What does calculating the yield tell you about a reaction?
8) Describe the five stages involved in producing chemicals in industry.
9) Give eight questions you should consider when deciding whether a process is sustainable.
10) What are the industrial conditions used for the Haber process?
11) What determines the choice of operating temperature for the Haber process?
12) What effect does the iron catalyst have on the reaction between nitrogen and hydrogen?
13) Explain why nitrogen fixation is important to world food production and industry.
14) Write the symbol equations for the three reactions in the Contact Process.
15) State and explain the conditions used in the Contact Process.
16) Why is there a limit to the concentration of ethanol that can be made using fermentation? How can the concentration be increased? Sketch a diagram of the apparatus you could use.
17) Describe how ethanol is made from crude oil. What conditions are needed?
18) Make a table to compare the sustainability of the three methods of ethanol production (fermentation of sugar, fermentation of waste biomass, and from ethane).
19) Describe in detail how you could prepare a pure sample of an ester.
20) Explain the difference between thermosoftening and thermosetting polymers.
21) What would you add to a polymer to make it stiffer and stronger?
22) Give an example of a "smart" material and describe how it behaves.
23) What are nanoparticles? Give two different applications of nanoparticles.
24) Why were CFCs initially popular?
25) What's the name for the part of the upper atmosphere that contains the ozone layer?
26) How are free radicals formed?
27) One CFC molecule can destroy thousands of ozone molecules. Why is this?

Section Twelve — Chemical Production

Index

Index

Index and Answers

Revision Summary for Section One (page 17)

5) Calcium

12) 14 H and 6 C

13)
H H H
| | |
H–C–C–C–H
| | |
H H H

14) a) $CaCO_3 + 2HCl \rightarrow CaCl_2 + H_2O + CO_2$

b) $Ca + 2H_2O \rightarrow Ca(OH)_2 + H_2$

Revision Summary for Section Three (page 37)

3) Propane — the fuel needs to be a gas at
−10 °C to work in a camping stove.

Revision Summary for Section Five (page 54)

2) b) 2 cm

c) 3.5 years

Page 67

A is simple molecular, B is giant metallic,
C is giant covalent, D is giant ionic

Revision Summary for Section Six (page 70)

7) 20.18

24) A: giant metallic,
B: giant covalent,
C: giant ionic

Page 71

1) Cu: 63.5, K: 39, Kr: 84, Cl: 35.5

2) NaOH: 40, Fe_2O_3: 160, C_6H_{14}: 86,
$Mg(NO_3)_2$: 148

Page 72

1) a) 30.0% b) 88.9% c) 48.0%
d) 65.3%

2) CH_4

Page 74

1) 21.4 g

2) 38.0 g

Revision Summary for Section Seven (page 77)

2) a) 40 b) 108 c) 44 d) 84 e) 106 f) 81
g) 56 h) 17

4) a) i) 12.0% ii) 27.3% iii) 75.0%
b) i) 74.2% ii) 70.0% iii) 52.9%

5) b) $MgSO_4$

7) 80.3 g

Revision Summary for Section Nine (page 100)

3) b)

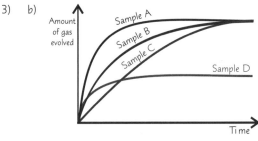

12) a) Mass of water heated = 116 g − 64 g = 52 g
Temperature rise of water = 47 °C − 17 °C = 30 °C
Mass of pentane burnt = 97.72 g − 97.37 g = 0.35 g

So 0.35 g of pentane provides enough energy
to heat up 52 g of water by 30 °C.

It takes 4.2 joules of energy to heat up 1 g of
water by 1 °C.

$Q = mc \, \Delta T$. Therefore, the energy produced in
this experiment is 52 × 4.2 × 30 = 6552 J.

So, 0.35 g of pentane produces 6552 J of
energy... meaning 1 g of pentane produces
6552/0.35 = 18 720 J or 18.720 kJ

Revision Summary for Section Ten (page 113)

10) R_f = 4.5 ÷ 12 = 0.375

15) a) No. of moles NaOH = 0.2 × (25 ÷ 1000) = 0.005
$HCl + NaOH \rightarrow NaCl + H_2O$,
so no. of moles HCl = 0.005
Concentration HCl (moles per dm³)
= 0.005 ÷ (49 ÷ 1000)
= 0.102 moles per dm³

b) M_r HCl = 1 + 35.5 = 36.5
mass = number of moles × M_r
= 0.102 × 36.5 = 3.72 grams per dm³

17) (0.479 ÷ 0.5) × 100 = 95.8%